Proceedings of the First European Astronomical Meeting
Athens, September 4-9, 1972

Volume 1

Solar Activity and Related Interplanetary and Terrestrial Phenomena

Edited by
J. Xanthakis

Springer-Verlag
Berlin · Heidelberg · New York 1973

With 78 Figures

ISBN-13: 978-3-642-65631-6 e-ISBN-13: 978-3-642-65629-3
DOI: 10.1007/978-3-642-65629-3

Softcover reprint of the hardcover 1st edition 1973

Contents

List of Contributors to Volume 1

BARBIERI, C., Astronomical Observatory, I-35100 Padova, Italy
VAN BEEK, H. F., Beneluxlaan 21, Utrecht, The Netherlands
BLAAUW, A., European Southern Observatory, D-2000 Hamburg 80, Bergedorfer Strasse 131
BONOV, A., University of Sofia, Department of Astronomy, Anton Avanov 1, Sofia 26, Bulgaria
BOTTER, R.,Centre d'Études Nucléaires de Saclay, Boite postal 2, F-91 Gif-sur-Yvette, France
BRÜCK, H. A., Royal Observatory, Edinburgh 9, Scotland, Great Britain
DELHAYE, J., INAG-Observatoire de Meudon, F-92 Meudon, France
ERIKSEN, G., Institute of Theoretical Astrophysics, P.O. Box 1029, Blindern, Oslo 3, Norway
FAHR, H. J., Institut für Astrophysik und Extraterrestrische Forschung der Universität Bonn, D-53 Bonn, Poppelsdorfer Allee 49, Germany
FALCIANI, R., Osservatorio Astronomico Capodimonte, Via Moiariello 16, I-80131 Napoli, Italy
DE FEITER, L. D., Astronomical Observatory Sonnenborgh, Utrecht, The Netherlands
FIREMAN, E. L., Smithsonian Institution, Astrophysical Observatory, Cambridge, MA 02138, USA
FOKKER, A. D., Sterrewacht Sonnenborgh, Servaasbolwerk 13, Utrecht, The Netherlands
FRICKE, W., Astronomisches Rechen-Institut, D-69 Heidelberg, Mönchhofstrasse 12–14, Germany
HOYNG, P., The Astronomical Institute, Space Research Laboratory, Utrecht, The Netherlands
DE JAGER, C., Space Research Laboratory, Beneluxlaan 21, Utrecht, The Netherlands
KIEPENHEUER, K. O., D-78 Freiburg, Schoeneckstrasse 6, Germany
KUBIČELA, A., Volgina 7, YU-11050 Belgrade, Yugoslavia
LANDINI, M., Osservatorio Astrofisico di Arcetri, Largo Fermi 5, I-50125 Firenze, Italy
MACRIS, C. J., Anagnostopoulou 14, Athens 136, Greece
MALTBY, P., Institute of Theoretical Astrophysics, P.O. Box 1029, Blindern, Oslo 3, Norway
MEZGER, P. G., M.P.I. für Radioastronomie, D-53 Bonn, Argelanderstrasse 3, Germany
MONSIGNORI FOSSI, B. C., Osservatorio Astrofisico di Arcetri, Largo Fermi 5, I-50125 Firenze, Italy
PALLAVICINI, R., Osservatorio Astrofisico di Arcetri, Largo Fermi 5, I-50125 Firenze, Italy
PAPAGIANNIS, M. D., Department of Astronomy, University of Athens, Athens, Greece
PETROPOULOS, B., 77 Kyprou, Athens 804, Greece
POULAKOS, C., Research Center for Astronomy and Applied Mathematics, Academy of Athens, Athens, Greece
PYYKKÖ, S. A., University of Oulu, Oulu, Finland
REPARIS, C. C., National Hellenic Research Foundation, Athens, Greece
RIGUTTI, M., Osservatorio Astronomico Capodimonte, Via Moiariello 16, I-80131 Napoli, Italy
STEVENS, G. A., The Astronomical Institute, Space Research Laboratory, Utrecht, The Netherlands
STRÖMGREN, B., Observatoriet, Østervoldgade 3, Copenhagen, Denmark
TRITAKIS, B., Research Center for Astronomy and Applied Mathematics, Academy of Athens, Athens, Greece
XANTHAKIS, J., Research Center for Astronomy and Applied Mathematics, Academy of Athens, 14 Anagnostopoulou, Athens 136, Greece
ZEREFOS, C. S., National Hellenic Research Foundation, Athens, Greece

Editor's Note

This volume contains the papers and communications presented at the First European Astronomical Meeting which, following a decision taken by the IAU, was held in Athens, Greece, from the 4th to 9th September 1972.

The papers and communications had been previously submitted to the Scientific Committee of the Meeting and had been duly approved.

Communications concerning European joint activities are likewise published at the end of the volume.

The Editor feels that it is his duty to thank the Greek Government for their financial support and also Springer-Verlag for publishing the volume and taking good care in its appearance.

J. Xanthakis
Editor

Address to the First European Astronomical Meeting

By H. E. C. A. PANAGHIOTAKIS
Hellene Minister of Culture and Sciences

Athens, 4th September, 1972

It is a great honour for a lay thinker to come to this meeting and bring to its participants a word of thanks and, if I may say so, encouragement in all the work they are doing or are about to do.

Perhaps no natural feature has such a firm place in the heart and mind of people as the sky. "This sky we have identified with the world" ("ὅν δέ οὐρανόν καί κόσμον ἐπωνομάκαμεν") Plato once said – and the Bible responded with the expression "the heavens proclaim the glory of God" („οἱ οὐρανοί διηγοῦνται δόξαν θεοῦ").

From the levels of music, poetry or geometry down to everyday life, the celestial influence on the stimulation of human imagination has been and is an immensely more potent force than earthly existence itself. The skies have taught to us the relativity not only of space or time but of experience as well. Something that we now see may no longer exist. The skies have opened to us the possibility of embracing many, very many, things of unlike substance and qualities which can exist together. The skies have shown us the endlessness of evolution and the proximity of extremes. It is unfortunate for mankind that political executives and administrators all over the world have not taken yet full advantage of these early discoveries of yours.

Achilles was one of the legendary heroes of Greece. He fought bravely against the ancient Trojans and gave his life in battle. I feel much indebted to Astronomy and to Lagrange in particular, because they proceeded to a posthumous reconciliation of the enemies, up in the skies. The first minor planet to provide a confirmation of the equilibrium conditions for several-body stability, the overall group of minor planets whose existence was predicted by mere reasoning, dictate through their detection and orbits in the solar system the principles of a new Iliad to be written, as I hope, by coming generations.

Honourable representatives and scientists from 34 countries, you have assembled into this land of Thales, Anaximandros, Anaximenes, Xenophanes, Parmenides, Zenon, Heracletos, Empedokles, Anaxagoras, Leykippos, Demokritos, Philolaos, Diogenes, Plato, Aristotle, Archimedes, Apollonios, Heron, and many others to discuss subjects ranging from celestial mechanics and wave propagation to stellar structure, chemistry and physics. Clearly and simply speaking, I wish I were enlisted among you. I can assure you that I was not born to become a Minister – yet, strange as it may sound, that there are still Ministers tempted by scientific knowledge or reasoning in any part of this solar planet.

Voicing on their behalf a warm message of goodwill and universal understanding, a message of tribute to the achievements of Astronomy and the devotion of astronomers, I wish that this First Meeting of its kind may be followed by a second and a third – and many more – in the years ahead.

Voicing on behalf of any layman or worker on this earth, who has carried out his own task in effort and in striving and who is entitled to expect a great deal from persons of your caliber in the near future, I declare this Meeting open.

Honourable participants, the stand is Yours ...

Inaugural Ceremony

Monday, 4th September, 1972, 10 a. m.

The inaugural ceremony was held in the aula of the Pantios School of Political Sciences in the presence of the Minister of Culture and Sciences and representatives of the Greek Universities.

On the platform were the Minister, Mr. CONSTANTINE PANAGHIOTAKIS, the president of the IAU, Prof. B. STRÖMGREN, the chairman of the Scientific Committee of the Meetings, Prof. A. BLAAUW, and the chairman of the local Committee, Prof. J. XANTHAKIS. After brief introductory addresses by Prof. XANTHAKIS, Prof. STRÖMGREN, and Prof. BLAAUW, the Minister of Culture and Sciences also addressed the meeting.

The Address of Professor J. Xanthakis

On behalf of the National Committee for Astronomy and the local Organizing Committee of these meetings I extend to you my warmest welcome to Athens.

It is an honor and a great pleasure for us to receive in the immediate vicinity of the Acropolis scientists of repute and also young astronomers from all over Europe.

It is well known that here, in this very locality, scientists and philosophers of antiquity, such as Pythagoras, Heraclitus, Philolaus, Aristotle and Aristarchus of Samos, discussed such fundamental problems as the shape of the earth, the distance and the motions of the Sun and planets, and in so doing gave birth to new ideas concerning the structure of the Universe, the Universe as it was then known.

The decision taken by the IAU, to organize on this very spot the first European meeting in order to discuss after twenty five centuries problems concerning the Sun and Galaxy, was, I believe, a very fortunate decision. Naturally, during these twenty-five centuries several of the problems which interested the ancient Greeks have found satisfactory solutions. New theories about the Universe have been developed, theories which make the ancient ideas appear naive and childish. I am afraid that our new theories will have the same fate and after twenty-five centuries, man, if he is lucky enough to survive, will, by travelling through space, be able to collect considerably more precise and significant information concerning the structure of the Universe.

The well-known saying by Socrates: "*γηράσκω ἀεί διδασκόμενος*" which means "the older I get the wiser I become" has not lost its validity during the twenty-five centuries which have elapsed.

Before closing this brief address I would like to ask you to be lenient about any possible shortcomings in the organization of these meetings; please bear in mind that the Committee had very little time at its disposal and has done its best.

I would like on behalf of the National Committee and of the organizing Committee to thank Greek Covernment for its financial support and also to thank the Bureau of the IAU for choosing the city of Athens to hold this first European meeting.

I wish you, ladies and gentlemen, a pleasant stay in this city of Pericles, Socrates and Plato, as well as every success in your scientific activities.

Indroductory Remarks

A need for closer contacts between the astronomers in Europe has been increasingly felt in the course of the past years. This need is well understood by anyone who is aware of the necessity of internationalization of research on our continent. Progress in astronomical research requires pooling of manpower and financial means in order that efforts in Europe may lead to achievements comparable to those elsewhere, particularly to those in the United States. Typical for this development are such international projects as ESRO, ESO, and possibly in the near future, JOSO and CESRA. At the same time, we observe a trend for internationalization of large national projects in both radio astronomy and optical astronomy.

Whereas in these large projects specially oriented groups have already arrived at an encouraging degree of integration across the national borders in Europe, it is felt by many an astronomer that the relations with colleagues in other countries outside these large projects still leave much to be desired. In most of the astronomical institutes in the European countries there is relatively little acquaintance with the research program on a small and intermediate scale in progress in the institutes of other countries. The language barrier undoubtedly is an important cause of this regrettable situation, but it would not seem to be the only one. Equally important seems to be the circumstance that there is little opportunity, especially for the younger, not yet internationally recognized, astronomers, to have contacts with their collagues of other countries outside the occasional symposia and colloquia. With the tendency for restricted participation, which is desirable in order to render such meetings fruitful for a follow-up by participants, they certainly do not suffice for the establishment of good contacts between the coming generation of astronomers.

Whereas in the United States the American Astronomical Society plays an essential role in promoting such contacts, such a common organization is lacking on the European continent. Therefore, a most important step towards remedying this situation was the initiative of the International Astronomical Union to create the institute of Regional Meetings. This initiative was supplemented admirably by the Greek National Committee for Astronomy, when early in 1972 it proposed to the IAU to hold a first European regional meeting in Athens. The present three volumes containing the proceedings of these meetings represent the fruit of this excellent effort.

For those who were present at the meetings in Athens, the memory will remain of an inspiring series of lectures and discussions, many of these latter very lively, and with that participation from astronomers from all over Europe (and also from across the ocean) which the International Astronomical Union's Executive

Committee must have had in mind when it initiated these meetings. Hopefully, it may induce colleagues from other states in Europe to organize similar meetings in the future. For the Scientific Organizing Committee, it was a privilege to be able to share in this first "regional" effort. It is a pleasure, on behalf of this committee, to acknowledge the excellent help it has had from the part of the Greek National Committee.

A. BLAAUW
Chairman,
Scientific Organizing Committee

Opening Words

By A. BLAAUW

Chairman, Scientific Organizing Committee

Ladies and Gentlemen!

On behalf of the Scientific Organizing Committee, I welcome the opportunity to say a few words on the program of this meeting. Early this year, after the Executive Committee of the International Astronomical Union and our Greek hosts, the Greek National Committee for Astronomy, had decided to organize this regional meeting, the task of formulating a suitable scientific program was delegated to our committee. Astronomers from many countries are represented on it. We were glad to assist in this way, notwithstanding the short terms on which the work had to be accomplished, because we all realized the importance of this event. We felt that we should be guided by the question, which kind of program might best promote the study of astronomy in Europe, and we oriented ourselves on the interests and developments of astronomy in most of the European countries. There was, of course, the difficult problem of getting many qualified speakers with relatively little time for preparation.

The Committee is pleased that it can offer the program as it is now in your hands. Many distinguished astronomers will contribute, either through their introductory lectures, or through subsequent papers and discussions. We hope we shall succeed in confronting you with accounts on *certain* developments typical for research in Europe. At an early stage we wondered whether to try to more or less evenly devote our time to most of the important domains of astronomy, or whether we should rather emphasize certain areas. We have preferred the second alternative in the optimistic anticipation of later regional meetings, elsewhere in Europe, which would offer opportunity for emphasizing other equally important fields.

If you scan through this program, you will note that fields emphasized mostly are those of galactic structure and the structure and evolution of other stellar systems; we have also chosen to emphasize the fastly developing field of infrared astronomy which is of great importance for both observational and for theoretical work. One afternoon session, tomorrow, will be devoted to a review of the large international projects which are in progress in Europe. Attention to these should be particularly fitting for our predominantly European audience. The Organizing Committee hopes that through that session we may stimulate the interest of European astronomers in these large projects as they may be expected to deeply influence the future progress in European astronomy, particularly in observational work.

On behalf of the Organizing Committee, may I already at this moment express our appreciation for the initiative and hospitality shown by the Greek National

Committee, and the support by the Ministry of Science, and express our gratefulness for the efficient help which the Program Committee has had from the part of the Assistant Secretary of the International Astronomical Union, Professor Contopoulos.

I wish you all a very profitable stay in Athens.

Opening Words

By B. STRÖMGREN

President, International Astronomical Union

Your Excellency Mr. PANAGHIOTAKIS

Professor XANTHAKIS

Ladies and Gentlemen

On the occasion of the Inaugural Ceremony of the First European Astronomical Meeting sponsored by the International Astronomical Union I am happy to have an opportunity to speak on behalf of the Union.

This Meeting is the first in what is hoped to be a series in IAU Sponsored Regional Meetings.

In the life of the Union the General Assemblies held at three-year intervals have traditionally been of great importance, and they still are and presumably will continue to be very important IAU events in spite of the difficulties connected with the organization of Assemblies of great size.

During the last several years the IAU sponsored Symposia and Colloquia have come to play a very great role. However, in order to be successful and fulfil their purpose these meetings – topical in character – have to have quite limited attendance.

For some time the need for another type of IAU sponsored Meeting has therefore been strongly felt. The Executive Committee of the IAU working closely with IAU National Committees and having the benefit of advice from a number of Astronomers in many countries a year ago formulated the rules for a new institution, the IAU Sponsored Regional Meetings. The character of these meetings would be different from Symposia and Colloquia in that a number of topics would be discussed, and in particular in giving many more astronomers an opportunity to participate.

The initiative for holding an IAU sponsored Regional Meeting is taken by a local Committee. In connection with IAU sponsorship the IAU appoints a Scientific Organizing Committee and the two Committees organize the Regional Meeting. The financial contribution of the IAU to a Regional Meeting is small and success of such a Meeting will depend to a large extent upon the efforts of the local Organizers.

We are most grateful to our colleagues and to the authorities in Greece for taking the initiative in calling and organizing this First European Astronomical Meeting and for having made such a great contribution toward making it a large and important Meeting. There is every expectation that this Meeting will be a very auspicious beginning and thus come to play an important role also in the life of the IAU.

On behalf of the International Astronomical Union I express our gratitude to our hosts at this First European Astronomical Meeting and extend all good wishes for the success of the Meeting.

Interplanetary Solar Phenomena

By L. D. de Feiter
The Astronomical Institute, Utrecht, The Netherlands

With 7 Figures

Abstract

In this review special emphasis has been given to those interplanetary phenomena that are of specific relevance to the physics of the solar corona. The topics discussed are; thermodynamic solar wind characteristics, solar wind composition, interplanetary magnetic field and solar active regions related to interplanetary structures.

I. Introduction

The basis for all studies in the field of Solar-Terrestrial Physics is the assumption that some terrestrial phenomena, like e.g. the occurrence of a strong auroral display, can be related in an unambiguous way to solar phenomena, such as the occurrence of a big flare a day or so before. Experience has shown that, although statistical relations of this kind really do exist, it is by no means an easy task to trace back in individual cases the solar origin of such terrestrial phenomena. It should be noted that in this case statistics are not of much help, but we should look carefully into the physical conditions prevailing on the sun, in interplanetary space as well as in the magnetosphere. The conditions prevailing in interplanetary space are of particular importance, since modifications of the "solar signals" may occur there and even new signals of purely interplanetary origin may originate, thus leading to false conclusions.

Bearing these critical remarks in mind and considering that the entire field of Solar-Terrestrial Physics has been recently reviewed during a specialized symposium in Leningrad (Dyer, 1972), I have interpreted my assignment in such a way that I will only discuss some of the solar and interplanetary aspects involved in Solar-Terrestrial Relations. The general scope of this review then becomes: "What do we know about the corona at 1 A.U. and how is it related to the corona at 1 $R_\odot$?" According to Evans (1971) we are in a fair way to understand the former better than the latter, so let us try and find out.

One of the limitations of our approach should be mentioned straight away: the observations we have to deal with are seriously limited in spatial coverage, and this limitation is most serious with respect to variations with heliographic latitude. The instruments probing the interplanetary medium remain almost strictly in the ecliptic plane and it is only by virtue of the inclination of 7° of the heliographic equator with respect to this plane that sampling in a small latitude band is provided.

This band, however, does not contain the solar activity belts and therefore future off-ecliptic missions will be needed in order to study, from a more appropriate viewing angle, the characteristics of the solar wind coming from these activity belts.

II. Thermodynamic Solar Wind Characteristics

Historically the idea of a tenuous gas streaming continuously away from the sun goes back to BIERMAN'S (1951) study of comet tails. Subsequently CHAPMAN (1954, 1957) showed that the inclusion of thermal conductivity in the equations describing a static corona, results in a model for the corona in which the temperature fall-off with distance is extremely slow. On basis of this work CHAPMAN concluded that the corona should extend even beyond the orbit of the earth. An unacceptable characteristic of CHAPMAN'S model, namely the very high coronal pressure at large distances, lead PARKER (1958) to the conclusion that a static corona could not be contained by the pressure of the interstellar medium and thus should expand. He then developed his hydrodynamic model of the corona which can be characterized by a continuous expansion with velocities reaching supersonic values at distances beyond a few solar radii. It is well-known that plasma observations in space have provided abundant support for PARKER'S ideas. The average quiet-time solar wind parameters as derived from an enormous amount of data gathered during the last ten years have been summarized in Table 1 (HUNDHAUSEN, 1970). It should be noted that the values in Table 1 are representative average values and that direct measurements reveal a large amount of variability (see Fig. 1). Some of these structures will be discussed in subsequent sections.

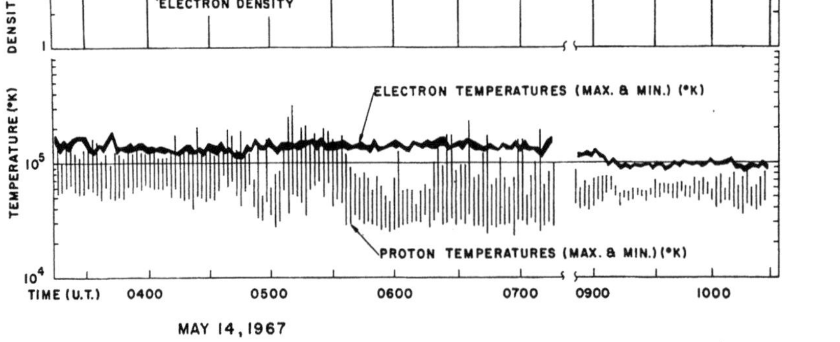

Fig. 1. Solar wind plasma properties observed on Vela 4B (after MONTGOMERY *et al.*, 1968)

Table 1. General quiet-time solar wind parameters (after HUNDHAUSEN, 1970)

Flow speed	320 km s^{-1}
Density	8 cm^{-3}
Total mass loss	1.2×10^{12} g s^{-1}
Proton temperature	4×10^4 K
Proton thermal anisotropy ratio	2
Electron temperature	1–1.5×10^5 K
Magnetic field strength	5×10^{-5} G
Kinetic energy flux	0.22 erg cm^{-2} s^{-1}
Heat conduction flux	0.01 erg cm^{-2} s^{-1}
Kinetic energy density	7×10^{-9} erg cm^{-3}
Proton thermal energy density	6×10^{-11} erg cm^{-3}
Electron thermal energy density	$\sim 1.5 \times 10^{-10}$ erg cm^{-3}
Magnetic field energy density	10^{-10} erg cm^{-3}

A first question to ask now is: "Do we understand these solar wind properties in terms of a realistic model for the outer solar corona?" In his introduction to a discussion on the structure and dynamics of the solar corona, KUPERUS (1971) remarks: "One does not expect a medium that emits so irregularly in such a broad spectral range to be described by simple physics; neither does one have much hope of being able to represent the coronal plasma by a few simple-state parameters." This remark can, almost unchanged, be applied to the solar wind.

It is well known that PARKER's hydrodynamic equations allow two types of solutions, referred to as solar wind (PARKER, 1958) and solar breeze (CHAMBERLAIN, 1961) solutions, respectively. It is only the solar wind solution that provides a fast enough drop of pressure at large distances, and that is consistent with space observations. However, the thermal energy of a coronal particle is only half the gravitational binding energy. How then can that particle escape from the sun? In their model calculations NOBLE and SCARF (1963) drew the energy needed for escape, at the solar wind velocity, from the heat conducted outward from the corona. For a coronal temperature of 2×10^6 K the density and velocity given in Table 1 can be reproduced provided a rather low density at the base of the corona, $N_0 \sim 3 \times 10^7$ cm^{-3}, is assumed. Increasing N_0 results in a higher density and a lower velocity at 1 A.U.; because less energy per atom is provided by thermal conduction, the velocity becomes smaller and, in order to conserve the flow of mass, the density must increase (PARKER, 1971). A way of bringing the coronal temperature and density in better accordance with the solar wind speed and density at 1 A.U. is to increase the temperature beyond a few solar radii, e.g. by extended heating by waves from the sun (NOBLE and SCARF, 1963). Several other mechanisms for this extended heating have been proposed: plasma instabilities and a resulting reduction of the thermal conduction (FORSLUND, 1970), or the obstruction of thermal conduction by the almost transverse magnetic field at large distances (PARKER, 1964). HARTLE and BARNES (1970) have shown by formal calculations that the most sensitive region for additional heating of the solar wind is between 1 and 25 $R_\odot$; the physical mechanism of the additional heating, however, is as yet unexplained.

Thus far only the so-called one-fluid models have been considered. Because the thermal conductivity is provided predominantly by electron-electron collisions, while the electron-proton collisions are relatively rare, the solar wind plasma should be treated as two only loosely coupled fluids. The two-fluid models derived by STURROCK and HARTLE (1966) and by HARTLE and BARNES (1970) give about the right electron temperature, but their proton temperature is about an order of magnitude below the actually observed value. The electron and the proton gas thus seem to be coupled more strongly than can be explained by Coulomb collisions. In solving this problem "liberal doses of wave dissipation may also help to bring the theoretical model more in line with observations" (PARKER, 1971).

For a long time the so-called exospheric theories of the solar wind have been overshadowed by the apparent success of PARKER's hydrodynamic theory and its modifications. In the work of BRANDT and CASSINELLI (1966), JOCKERS (1970) and HOLLWEG (1970) it has been shown that several characteristics of the solar wind can be derived from this approach as well; however it does not seem possible to allow for the low value of the temperature anisotropies.

Although the general features of the relations between the conditions in the solar corona and those in the solar wind at 1 A.U. seem to be reasonably well understood, some unknown modifications in the region between the sun and the earth seem to exist. It is also important to note that the solar wind we observe near the earth's orbit might consist of a superposition of fibril-like streams originating in individual active regions, thus masking the properties of the quiet-time solar wind which we have been trying to relate to the quiet corona (PARKER, 1971).

III. Solar Wind Composition

In Fig. 2 a typical energy-per-charge spectrum for the solar wind plasma is presented (BAME *et al.*, 1968). Several species of ions have been identified in the figure. Apart from the proton peak, at about 0.8 keV, the most pronounced peak is due to helium at 1.6 keV. For several reasons I will restrict myself here to a discussion of the helium to hydrogen ratio and its variability with time. The average solar wind helium abundance, as determined from a large number of spectra, has been found to be 0.045. This value is somewhat lower than that derived using

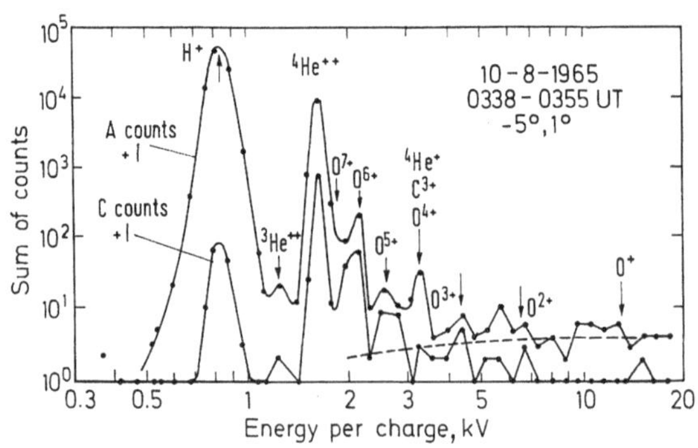

Fig. 2. Energy per charge spectrum observed with the analyser on Vela 3A. The two spectra marked with A and C counts refer to different pulse-size thresholds for the electron multiplier (after BAME *et al.*, 1968)

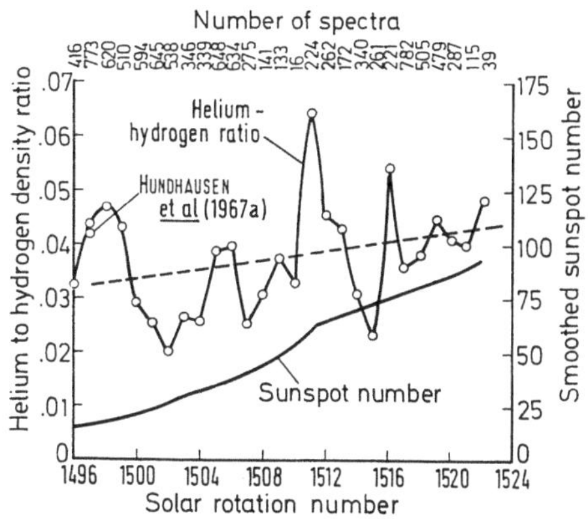

Fig. 3. Helium to hydrogen density ratio as averaged over 27 days between July 1965 and July 1967. Also indicated is the smoothed sunspot number (after Robbins *et al.*, 1970)

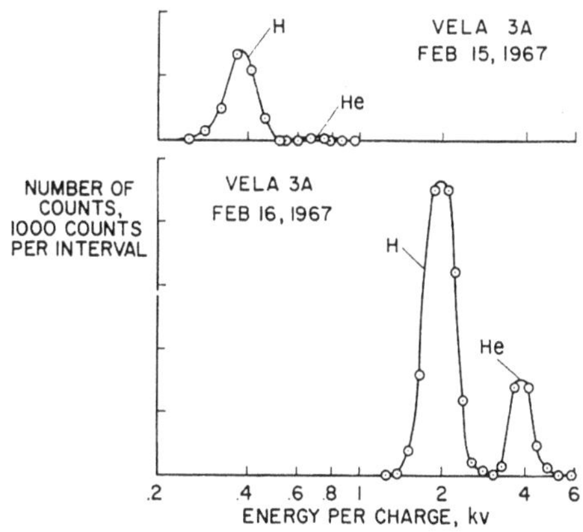

Fig. 4. Energy per charge spectra before and after the arrival of a helium-enriched flare plasma (after Hirshberg *et al.*, 1972)

other methods: 0.09 from theoretical considerations of the solar interior, 0.06–0.09 from abundance ratios in solar cosmic rays, and 0.05–0.06 from upper limits of the solar neutrino flux. Due to the uncertainties in these determinations no definite conclusions can be drawn from these differences.

The variability in time of the helium abundance, and its apparent relation with solar activity can be seen in Fig. 3 (Robbins *et al.*, 1970). A detailed study of this variability reveals that periods for which the 10-day average of the helium abundance exceeds 6% closely correlate with the presence on the solar disk of large active centers in which several flares occur. Also a good correlation has been found between a high helium abundance and geomagnetic variability (Robbins *et al.*, 1970). Very high helium abundances, up to 29% (Hirshberg *et al.*, 1972) have been found for the driver gas expelled during very important flares. Twelve well-documented events of this kind form the basis of a study by Hirshberg *et al.* (1972); an example is given in Fig. 4.

The interpretation of the highly variable helium content is not yet clear, although promising attempts have been made. Thermal gradients, both vertical, in the transition region, and horizontal, near active regions, produce local inhomogeneities in the ion composition; thus heavier elements migrate toward regions of high temperature (e.g. NAKADA, 1969). The work of GEISS *et al.* (1970) shows that in the lower region where the solar wind is formed, below a few $R_\odot$, the larger mass and the lower charge-to-mass ratio of helium ions results in lower velocities than those for the protons. At larger distances the velocities tend to become equal. Following this reasoning we may conclude that the normal solar wind has a lower helium abundance than the normal solar material, and leaves behind a layer where the helium content increases with time; this layer is expelled from the sun during active periods. It will be clear that this stratification cannot explain the apparent correlation with specific active regions and that an enhanced concentration above active regions seems therefore to be a necessary conclusion.

IV. The Interplanetary Magnetic Field

In the outer corona, where the acceleration to solar wind velocities takes place, the ratio between the gas-kinetic pressure and the magnetic pressure gradually increases until, at large distances, the magnetic morphology is completely determined by the radial streaming of the solar wind. Also in this region a gradual bending of the magnetic field lines toward the radial direction takes place. Because of the solar rotation these field lines, initially radially directed, become the well-known Archimedian spirals (PARKER, 1963).

Models have been developed by ALTSCHULER and NEWKIRK (1969) and by SCHATTEN (1969). They allow the determination of the general topology of the magnetic field in the corona and in interplanetary space, beginning from the observed distribution of the photospheric magnetic field. The assumptions underlying these model calculations are: force-free fields in the corona and a "source surface" where the field direction becomes radial through the dynamic pull of the solar wind, which is usually taken at a distance of 2.6 $R_\odot$. These models very well explain the gross characteristics, including the well-known sector structure (NESS and WILCOX, 1964). It is mostly the weak fields extending over large areas, that are drawn out into interplanetary space; the strong localized fields form closed loop-like structures that resist the dynamic solar wind pressure.

Assuming a radially streaming solar wind, a corotating field pattern and the transformation equations for the field components with distance as derived for a convected field, the observations at 1 A.U. can be used to derive the ecliptic magnetic field at various other distances. Part of such a map is reproduced in Fig. 5 (SCHATTEN *et al.*, 1968). This particular example has been chosen because it displays a loop-like structure of field lines that extends all the way to 1 A.U. It is an enlargement of a structure that existed in the corona some five days before.

An independent demonstration of the spiral structure can be derived from Fig. 6, where the locus of emission of an interplanetary type III burst extending nearly to the earth' orbit has been drawn. The assumption here is that a beam of

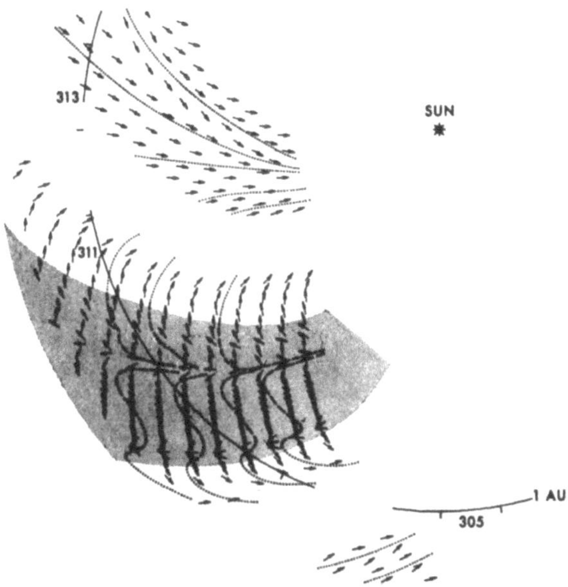

Fig. 5. Magnetic flux loop extending out to 1 A.U. (after SCHATTEN *et al.*, 1968)

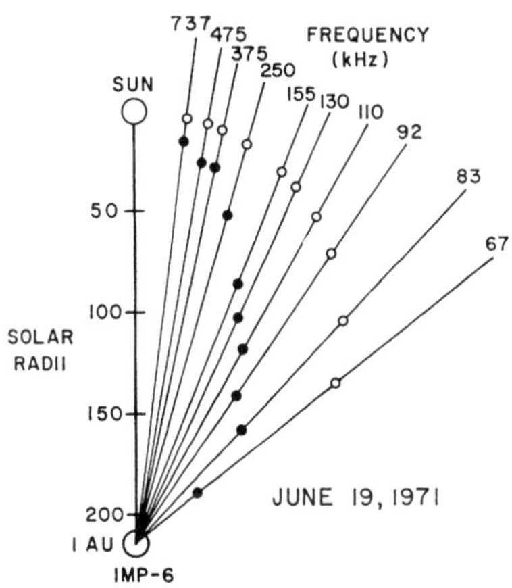

Fig. 6. Directions of arrival for interplanetary radio emission at frequencies of 67 kHz to 737 kHz. Points labelled by a dark dot have been derived using the assumption of emission at the local plasma frequency, those marked by open circles refer to emission regions located at the closest distances to the sun permitted by the direction of arrival. The nature of the spiral magnetic field is indicated by these observations (after FAINBERG *et al.*, 1972)

high-energy electrons is directed along the magnetic field, exciting radiowaves at the local plasma frequency (FAINBERG *et al.*, 1972).

The interplanetary magnetic field correlates very well with the mean solar magnetic field, which can be determined from the ZEEMAN splitting of spectral lines in sunlight integrated over the whole disk. In this measurement the "local"

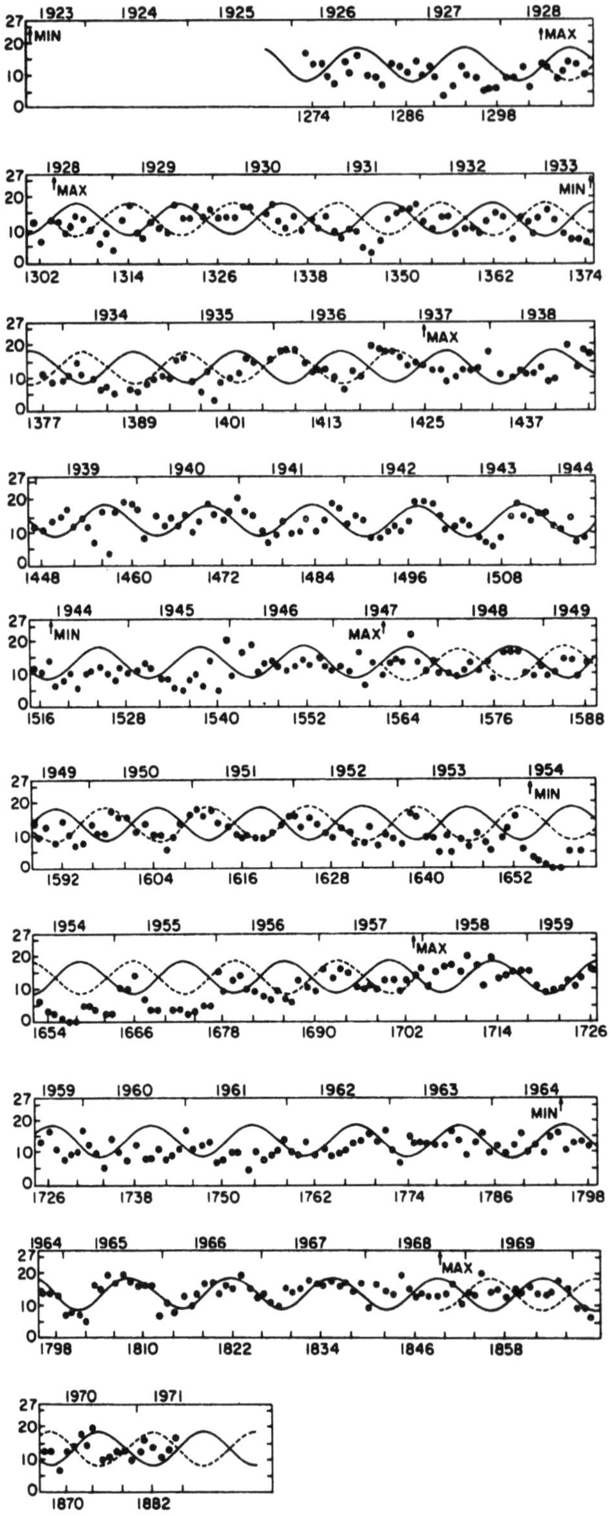

Fig. 7. The number of days, per 27-day Bartels' rotation period, with an inferred interplanetary magnetic field polarity directed toward the sun, for the period 1927–1971. The solid curve is the heliographic latitude of the earth; the dashed line is the same latitude curve shifted 180° (after WILCOX and SCHERRER, 1972)

strong fields of opposite polarity cancel and thus the main contribution comes from extended weak field regions which also have been shown to contribute to the interplanetary field.

Recently SVALGAARD (1972) discovered a method by which the polarity of the interplanetary magnetic field (towards or away from the sun) can be inferred from geomagnetic field variations at polar latitudes. Using data from the station Godhavn SVALGAARD (1972) was able to determine the polarity of the interplanetary magnetic field for the 45-year period 1926–1971. For this period WILCOX and SCHERRER (1972) determined for each BARTELS' rotation of 27 days the number of days with inward field polarity. Their result is reproduced in Fig. 7. A clear annual wave of this polarity parameter can be noted from this figure; it is stressed by the plot of the heliographic latitude of the earth which varies with an amplitude of 7°.25 and is zero on December 7 and June 7. This annual wave has been suggested, but then on the basis of only a few years' data, earlier by ROSENBERG and COLEMAN (1969). The phase of the annual wave with respect to heliographic latitude of the earth seems to change by 180° when going from on solar cycle to the next, thus indicating the solar magnetic cycle of about 20 years. The analysis of these observations is a fine example of the way in which solar physics can benefit from studies in the field of solar-terrestrial physics.

V. Interplanetary Structures Related to Specific Active Regions

As noted earlier in this review, the parameters characterizing the solar wind plasma show a high degree of variability with time scales ranging from the limit of resolution to several rotations. Already from the first series of observations the existence of recurrent high-velocity streams was noted by SNIJDER *et al.* (1963). Attempts to relate these streams to specific solar regions was not successful (SNIJDER and NEUGEBAUER, 1966). Because of the very good correlation existing between the solar wind speed and the geomagnetic K_p index, which was found by SNIJDER and NEUGEBAUER (1966), the search for the solar origin of high-velocity recurrent streams is a continuation of earlier investigations into the identification of the so-called *M*-regions. From a statistical study ROOSEN (1969) found that faint sources of microwave emission tend to be associated with periods of enhanced geomagnetic activity. COUTURIER and LEBLANC (1970), on the other hand, found a close association of high-velocity solar wind streams with radio storm centers at meter wave-lengths which also show an enhanced emission in the green coronal line (λ 5303). Qualitatively these results are in agreement with the perturbation models of CAROVILLANO and SISCOE (1969) and SISCOE and FINLEY (1969).

VI. Concluding Remarks

In the foregoing sections some of the recent work on interplanetary phenomena has been reviewed. As has been indicated in the introduction no full coverage of the entire field has been attempted, but rather a few illustrative subjects have been selected. For more information the reader is referred to the comprehensive reviews like: WILCOX (1968), HUNDHAUSEN (1970), MCCRACKEN and RAO (1970) and SCHATTEN (1971).

References

ALTSCHULER, M. D., NEWKIRK, G.: Solar Phys. **9**, 131 (1969).

BAME, S. J., HUNDHAUSEN, A. J., ASBRIDGE, J. R., STRONG, I. B.: Phys. Rev. Lett. **20**, 393 (1968).

BIERMANN, L.: Z. Astrophys. **29**, 274 (1951).

BRANDT, J. C., CASSINELLI, J. P.: Icarus. **5**, 47 (1966).

CAROVILLANO, R. L., SISCOE, G. L.: Solar Phys. **8**, 401 (1969).

CHAMBERLAIN, J. W.: Astrophys. J. **133**, 675 (1961).

CHAPMAN, S.: Astrophys. J. **120**, 151 (1954).

CHAPMAN, S.: Smithsonian Contrib. Astrophys. **2**, 1 (1957).

COUTURIER, P., LEBLANC, J.: Astron. Astrophys. **7**, 254 (1970).

DYER, E. R. (Ed.): Solar-terrestrial physics 1970. Dordrecht-Holland: D. Reidel 1972.

EVANS, J. W.: In: C. J. MACRIS (ed.), Physics of the solar corona, p. 1. Dordrecht-Holland: D. Reidel 1971.

FAINBERG, J., EVANS, L. G., STONE, R. G.: Science (to be published 1972).

FORSLUND, D. W.: J. Geophys. Res. **75**, 17 (1970).

GEISS, J., HIRT, P., LEUTWIJLER, H.: Solar Phys. **12**, 458 (1970).

HARTLE, R. E., BARNES, A.: J. Geophys. Res. **75**, 6915 (1970).

HIRSHBERG, J., BAME, S. J., ROBBINS, D. E.: Solar Phys. **23**, 467 (1972).

HOLLWEG, J. V.: J. Geophys. Res. **75**, 2403 (1970).

HUNDHAUSEN, A. J.: Rev. Geophys. Space Phys. **8**, 729 (1970).

JOCKERS, K.: Astron. Astrophys. **6**, 219 (1970).

KUPERUS, M.: In: C. DE JAGER (ed.), Solar-terrestrial physics 1970. Part I—The sun. Dordrecht-Holland: D. Reidel 1971.

MCCRACKEN, K. G., RAO, U. R.: Space Sci. Rev. **11**, 155 (1970).

MONTGOMERY, M. D., BAME, S. J., HUNDHAUSEN, A. J.: J. Geophys. Res. **73**, 4999 (1968).

NAKADA, M. P.: Solar Phys. **7**, 302 (1969).

NOBLE, L. M., SCARF, F. L.: Astrophys. J. **138**, 1169 (1968).

PARKER, E. N.: Astrophys. J. **128**, 664 (1958).

PARKER, E. N.: Interplanetary dynamical processes. New York: Interscience 1963.

PARKER, E. N.: Astrophys. J. **139**, 93 (1964).

PARKER, E. N.: Rev. Geophys. Space Phys. **9**, 825 (1971).

ROBBINS, D. E., HUNDHAUSEN, A. J., BAME, S. J.: J. Geophys. Res. **75**, 1178 (1970).

ROOSEN, J.: Solar Phys. **8**, 450 (1969).

ROSENBERG, R. L., COLEMAN, P. J.: J. Geophys. Res. **74**, 5611 (1969).

SCHATTEN, K. H.: Rev. Geophys. Space Phys. **9**, 773 (1971).

SCHATTEN, K. H., NESS, N. F., WILCOX, J. M.: Solar Phys. **5**, 240 (1968).

SCHATTEN, K. H., NESS, N. F., WILCOX, J. M.: Solar Phys. **9**, 442 (1969).

SISCOE, G. L., FINLEY, L. T.: Solar Phys. **9**, 452 (1969).

SNIJDER, C. W., NEUGEBAUER, M.: In: R. W. MACKIN and M. NEUGEBAUER (eds.), The solar wind, New York: Pergamon 1966.

SNIJDER, C. W., NEUGEBAUER, M., RAO, U. R.: J. Geophys. Res. **68**, 6361 (1963).

STURROCK, P. A., HARTLE, R. E.: Phys. Rev. Lett. **16**, 628 (1966).

SVALGAARD, L.: Geophys. Pap. R-6, Danish Meteor. Inst., Charlottenlund, Denmark (1968).

SVALGAARD, L.: J. Geophys. Res. **77**, 4027 (1972).

WILCOX, J. M.: Space Science Rev. **8**, 258 (1968).

WILCOX, J. M., SCHERRER, P. H.: J. Geophys. Res. **77**, 5385 (1972).

Discussion

HOUTGAST, J.:
Would you explain what you mean with the expression: rigid rotation of the interplanetary magnetic field vectors?

DE FEITER, L. D.:
The rigid rotation cannot extend beyond a few solar radii. Nevertheless one sees these recurrence after 27 days. This should be interpreted as the effect of a continous enhanced outflow of the region concerned, the effects of which, observed at I.A.U. should also recur.

FALCIANI, R.:
Do you know the existence of solar neutrons observations?

DE FEITER, L. D.:
There exist as far as I know no positive neutron observations. As you may know one should also look for γ-ray lines around a few MeV. I hope that Dr. CHUBB's instrument on board OSO VII will provide observation on the flares of the beginning of August this year. Dr. SVESTKA's and my ideas are that it is only the very powerful white-light flares that should give such emissions. After all one needs a sizeable number of collisions to produce these radiations.

FAHR, H. J.:
I remember having seen a publication by MIHALOV and WOLFE where they analyse Pioneer VI and VII data. These show a slight anti-correlation between proton activities and solar wind velocities under disturbed conditions. This seems to be in contrast to what you pointed out.

DE FEITER L. D.:
You may refer to my picture on the proton density as a function of time. You are right that the proton densities itself are anti-correlated with the solar wind speed. In Fig. 1 the proton velocity is almost constant.

Solar Flares During the Past 1000 Years as Revealed by Lunar Studies

By E. L. Fireman
Smithsonian Astrophysical Observatory, Cambridge, Massachusetts, USA.

With 3 Figures

Abstract

The intensities of the >50-MeV proton fluxes in several recent solar flares are obtained from measurements of Ar^{37} radioactivity as a function of depth in lunar material. These intensities agree within a factor of 2 with measurements made by detectors aboard satellites. The intensities of the solar proton fluxes of >200-Mev and >50-Mev energy averaged over the past 1000 years are estimated from measurements of Ar^{39} radioactivity as a function of depth in lunar material. The intensity of >200-Mev solar-flare protons obtained from Ar^{39} in Fe, when averaged over the past 1000 years, was equal to that of galactic cosmic rays. This result indicates that the 19th solar cycle, which had a number of very energetic solar flares, was more typical of solar activity during the past 1000 years than was the 20th. The intensity of >50-Mev solar-flare protons obtained from Ar^{39} in Ca was $8 \times 10^7/cm^2$ year averaged over the past 1000 years, which corresponds to three flares per year with intensities equal to the flare of January 24, 1971.

Introduction

Solar flares and galactic cosmic-ray particles produce radioactivities in lunar material. Large sudden increases in particle flux are associated with solar flares. Galactic cosmic rays are relatively constant. Solar flares also contain a much larger abundance of particles of 10- to 200-Mev energy relative to higher energy particles than do cosmic rays. Because of particle attenuation and multiplication in the material, the amounts of radioactivities produced depend on the depth of the lunar material. The amount of radioactivity at a given depth, when its production cross-section is known, is a monitor of the solar-flare and cosmic-ray flux. A radioactivity integrates the flux over a time corresponding to approximately three half-lives. Radioactivities with short half-lives (1 month or less) monitor recent solar flares; those with long half-lives (greater than 100 years) monitor ancient solar flares. Recent solar flares have also been monitored by electronic detectors aboard satellites. The results of these two completely different types of monitors should be compared. If the results agree reasonably well, then from the long-lived radioactivities one can confidently make deductions about ancient solar flares.

This work was supported in part by contract NGR 09-015-145 from the National Aeronautics, and Space Administration.

Lunar material has been collected by five Apollo missions: two occurred within 1 month after the large solar flares of November 2, 1969, and January 24, 1971, respectively; for two others no significant solar flares occurred 4 or more months before the missions; and for one only a minor flare occurred during the mission. We obtained material from known depths collected from four of these missions. We have measured Ar^{37} (35-day half-life), Ar^{39} (270-year half-life), and H^3 (12.4-year half-life) in some of these samples. The Ar^{37} measurements give intensities for the solar flares of November 2, 1969, January 24, 1971, and April 19, 1972. The Ar^{39} measurements give information about solar-flare intensities averaged over the past 1000 years; for this period some information about solar activity from sunspot numbers is available and some information about cosmic rays from C^{14} data and meteorites exists. The tritium covers a time (~ 50 years) for which there are measurements on both cosmic rays and solar flares, but the relation between tritium and solar flares will not be discussed here. Measurements by others (FINKEL *et al.*, 1971; RANCITELLI *et al.*, 1971; KELLEY *et al.*, 1971) on the Na^{22} (2.6-year half-life) and Al^{26} (0.74×10^6-year half-life) activities show that the solar particles (>10-Mev energy) averaged over the past 10 years and 2×10^6 years have had roughly similar intensities. The Ar^{37} and Ar^{39} activities should give more detailed information about solar activity over the past 1000 years. From various measurements on solar activity since 1950 it is known that the 19th solar cycle (1950–1961) had larger and more energetic solar flares than did the 20th solar cycle (1961–1972). An interesting question to be answered from the argon activities is which of these solar cycles is more typical of solar activity during the past 1000 years.

Samples

The important lunar samples for our studies are documented ones, which means they are from known depths. Some documented samples are soil samples taken from known locations in a core obtained by pressing a tube into the lunar surface. Fig. 1 is a photograph of a core tube after it was cut open at the Manned Spacecraft Center in Houston, Texas. We obtained soil samples of 1 to 2 g from four known locations in such a tube from the Apollo 16 mission; the depths were ~ 1.0, ~ 6.0, ~ 15, and ~ 26 g/cm^2. We obtained samples from similar depths in documented rocks photographed *in situ* before they were picked up by the astronauts and delivered to Houston. These rocks were 12002 from Apollo 12, 14321 from Apollo 14, and 15555 from Apollo 15. Fig. 2 is a photograph of rock 15555 *in situ* on the moon. Sample 15555 is a crystalline rock approximately 14 cm high; it has often been termed football-sized and is the largest studied.

Experimental Procedure

We have already described our experimental method in detail (FIREMAN *et al.*, 1970; D'AMICO *et al.*, 1971). Briefly, we melt the sample by induction heating in the presence of argon and hydrogen carriers, remove the gases from the molten sample, purify the argon and hydrogen in the gases, transfer these purified com-

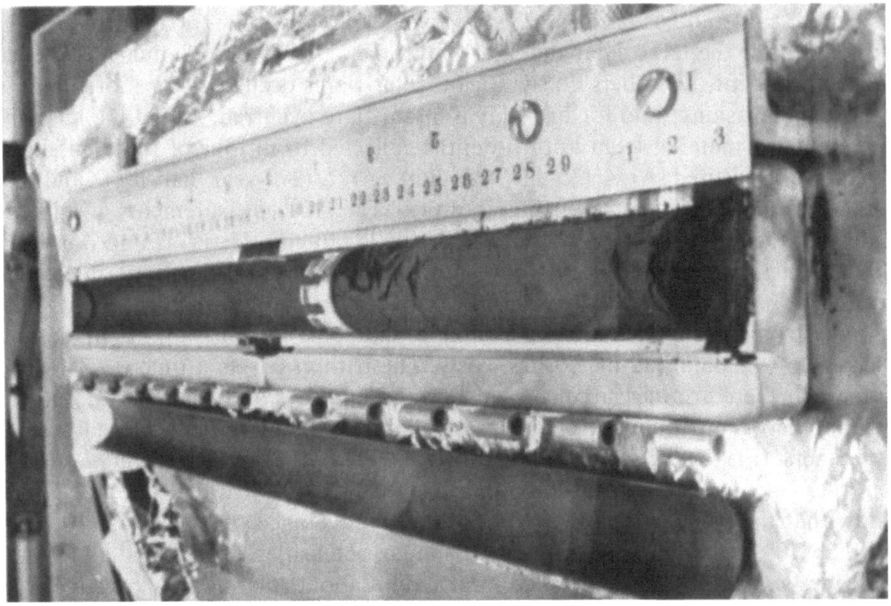

Fig. 1. Core tube containing documented lunar soil, after being sliced open

ponents into separate low-level proportional gas counters, and count their activities. The purification elements are finely divided vanadium powder and titanium pellets. After purification the volume of the argon is measured to determine yield. The volume is always between 95 and 100% that of argon carrier. The volume of the argon counter is 0.5 cc, and the argon pressure in the counter is approximately 1.5 atm. Methane (10%) is added to the argon in the counter for good counting characteristics. The counts of energy between 0.4 and 8.0 keV are recorded on a 100-channel analyzer, and those above 7.4 kev, on scalers of the low-level counting system. The Ar^{37} appears as a peak in the counting data at 2.8 kev, with a 0.6-kev half-width. Representative data from lunar samples are given in previous publications (FIREMAN *et al.*, 1970; D'AMICO *et al.*, 1971).

The Intensities of Recent Solar Flares

Bombarding targets of simulated lunar material (D'AMICO *et al.*, 1971) and targets of pure elements (STOENNER *et al.*, 1971) by protons of various energies in cyclotrons shows that Ar^{37} in lunar samples is produced by the action of protons of energy $\geqq 50$ Mev. The principal nuclear reaction for the Ar^{37} production in lunar material is the $Ca^{40}(n, \alpha)Ar^{37}$ reaction; the neutrons that cause this reaction to proceed with high probability are produced by the action of energetic protons ($\geqq 50$ Mev) on lunar material.

Table 1 gives our Ar^{37} results for samples with known depths in documented material from the Apollo 12, 14, 15, and 16 missions. The Ca contents as deter-

Table 1. Ar^{37} activities, depths, and Ca contents

Sample	Depth (g/cm^2)	Ar^{37} (dpm/kg)	Ca (%)	Ar^{37}/Ca (dpm/kg Ca)
60007.93, 94, 95	0–2.0	39.0 ± 2.2	11.4	352 ± 20
60007.96, 97, 98	5.0–7.0	47.9 ± 1.9	11.4	422 ± 17
60007.99, 100, 101	14.0–16.0	58.5 ± 2.4	11.4	513 ± 22
60007.102, 103, 104	25.3–27.3	68.1 ± 2.9	11.4	598 ± 25
15555.98	0–2.0	15.8 ± 2.8	6.7	236 ± 42
15555.80	~15	31.2 ± 3.3	6.7	465 ± 49
15555.77	~40	35.0 ± 6.5	6.7	520 ± 97
15271.17	~ 2	21.4 ± 2.0	8.1	264 ± 25
14321.81, 83	0–1.5	38.5 ± 2.5	6.1	632 ± 39
14321.81, 83	1.5–3.0	37.6 ± 2.5	6.1	618 ± 39
14321.81, 83	3.0–4.5	35.0 ± 4.0	6.1	574 ± 63
14321.95	~35	34.0 ± 2.7	6.1	537 ± 42
12002.57	0–2.4	30.0 ± 4.0	5.4	556 ± 65
12002.57	2.5–9.0	25.0 ± 1.5	5.4	463 ± 28
12002.59	14.5–19.0	27.5 ± 2.5	5.4	510 ± 46

mined by other investigators are also given. Only the Apollo 15 material was not subjected to any solar flares for 4 months before recovery. Near-surface samples from Apollo 15 material had the lowest Ar^{37}/Ca ratios. The top sample of rock 15555, which was from a depth 0 to 2 g/cm^2, and soil sample 15271.17, which had been scooped from the lunar surface, had Ar^{37}/Ca ratios of 236 ± 42 and 264 ± 25 dpm/kg Ca, respectively. On the other hand, the top sample of rock 12002 and the top soil sample from core 60007 had Ar^{37}/Ca ratios of 632 ± 39 and 352 ± 20 dpm/kg Ca, respectively, at the time of recovery. The Ar^{37}/Ca ratios in rock 15555 are caused solely by galactic cosmic rays. The excess Ar^{37}/Ca in other samples over that observed at similar depths in 15555 results from solar-flare particles. The difference between the Ar^{37}/Ca in any one of the samples and the Ar^{37}/Ca at the same depth of 15555 has been corrected for decay of Ar^{37} activity during the time between the date of the last known solar flare and the date of sample recovery. The corrected difference, ΔAr^{37}/Ca, is given in Table 2. Near the top of the lunar surface, the flare of April 19, 1972, produced a ΔAr^{37}/Ca of 98 ± 38 dpm/kg Ca; that of January 24, 1971, 490 ± 52 dpm/kg Ca; and that of November 2, 1969, 455 ± 105 dpm/kg Ca. To obtain solar-flare intensities from these values, it is necessary to know the Ar^{37} production from Ca as a function of proton energy and the solar-flare proton-energy spectrum. D'Amico *et al.* (1971) measured the Ar^{37} production cross-section from Ca in targets of simulated lunar material to be 60 ± 3 mb for protons 80 and 160 Mev and 10.8 ± 1.0 mb for those of 50 Mev. On the basis of these target measurements, we have assumed that the Ar^{37} production cross-section rises linearly from 0 for 44-Mev protons to 60 mb for 80-Mev protons but is constant and equal to 60 mb for proton energies above 80 Mev. We have also assumed a differential energy spectrum of $1/E^3$ for the solar-flare protons. With these cross-section values and this energy spectrum, the solar-flare proton fluxes that result from the ΔAr^{37}/Ca values are

(a)

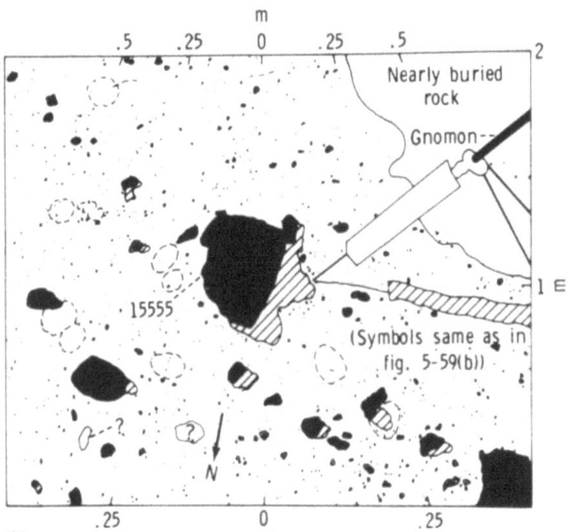

(b)

Fig. 2. Rock 15555, a football-sized lunar rock *in situ* on the moon and in documentation drawing

given in Table 2. For the flare of November 2, 1969, we obtain a flux of $(3.4 \pm 0.9) \times 10^7/cm^2$; for that of January 24, 1971, $(3.7 \pm 0.6) \times 10^7/cm^2$; and for that of April 19, 1972, $(0.7 \pm 0.3) \times 10^7/cm^2$. If we had assumed a steeper differential energy spectrum, we would have obtained lower flux values. With data from detectors aboard a satellite, J. VAN ALLEN (private communication, 1971) obtained $(3.1 \pm 0.6) \times 10^7/cm^2$ and $(2.0 \pm 0.4) \times 10^7/cm^2$ for the fluxes of protons >55 Mev

Table 2. Solar-flare proton fluxes (≧50 Mev) from Ar^{37}/Ca and from satellite data

Flare date	$\Delta Ar^{37}/Ca$ (dpm/kg Ca)	Time before recovery (days)	$(\Delta Ar^{37}/Ca)_{cor.}$ [a] (dpm/kg Ca)	Proton flux[b] ($10^7/cm^2$)	Proton flux[c] ($10^7/cm^2$)	Proton flux[d] ($10^7/cm^2$)
April 19, 1972	95 ± 35	2	98 ± 38	0.7 ± 0.3	–	–
January 24, 1971	370 ± 40	13	490 ± 52	3.7 ± 0.6	3.1 ± 0.6	2.9
November 2, 1969	300 ± 70	21	455 ± 105	3.4 ± 0.9	2.0 ± 0.4	2.0

[a] Corrected for elapsed time between flare and sample recovery.
[b] Integral flux (≧50 Mev) over 2π solid angle from $\Delta Ar^{37}/Ca$ and cross-sections from D'AMICO *et al.* (1971).
[c] Integral flux (>55 Mev) over 2π solid angle (private communication from J. VAN ALLEN, 1971).
[d] Integral flux (>60 Mev) over 2π solid angle BOSTROM (1971).

for the flares of November 2, 1969, and January 24, 1971, respectively. From similar detectors (BOSTROM, 1971) aboard another satellite, C. BOSTROM obtained similar values for the solar protons >60 Mev in these flares. Our flux values from $\Delta Ar^{37}/Ca$ are in reasonable agreement with theirs. The agreement indicates that the determination of the solar-flare flux integrated over long time spans obtained from long-lived radioactivities should be reliable when the contribution from galactic cosmic rays can be reliably estimated.

Fig. 3 shows plots of the Ar^{37}/Ca ratio as a function of depth for the material from the four Apollo missions. At large depths (≧15 g/cm²), the Ar^{37}/Ca ratios from different missions are the same. This indicates that none of the recent solar flares had significant numbers of protons above 300-Mev energy. All the Ar^{37} at the large depths are attributed to galactic cosmic rays.

The Solar-flare Intensity Averaged over the past 1000 Years

From cross-section measurements on Ar^{39} (270-year half-life) production in various elements mainly by STOENNER *et al.* (1971) and from Ar^{39} measurements in separated lunar fractions of different chemical composition (BEGEMANN *et al.*, 1970), one can conclude that four target elements – Fe, Ti, Ca, and K – contribute to Ar^{39} production in lunar material. To extract information about solar flares averaged over the past 1000 years from Ar^{39} measurements in lunar material, it is necessary to devise a procedure by which the Ar^{39} in the four elements as a function of depth can be obtained. If the Ar^{39} activity is known in four lunar samples (same depth) with different Fe, Ti, Ca, and K abundances, the Ar^{39} in the separate elements can be calculated from four linear equations with four unknowns. Although we have measured Ar^{39} as a function of depth in four different lunar materials, among only three of them are the chemical differences sufficiently large to make this calculation useful. Rock 14321 had 15 times more K, and 60007 soil samples had twice as much Ca, as did rock 15555; however, between rocks

Table 3. Ar^{39} produced[a] in Fe, Ca, and K versus depth in the moon

Depth (g/cm^2)	Fe → Ar^{39} (dpm/kg Fe)	Ca → Ar^{39} (dpm/kg Ca)	K → Ar^{39} (dpm/kg K)
1.0	20 ± 4	60 ± 7	700 ± 200
~ 5.0	22 ± 3	50 ± 8	1000 ± 300
15.0	21 ± 4	35 ± 7	2000 ± 300
~30.0	22 ± 4	62 ± 12	2200 ± 300

[a] Ar^{39} production in Ti is assumed to be 4 times that in Fe; calculated from Ar^{39} data given in FIREMAN *et al.* (1972).

12002 and 15555, there was no large difference in chemical composition. We carried out the calculation for Ar^{39} in three separate elements – Fe, Ca, and K – as a function of depth by combining the Fe and Ti contributions. We assumed that the Ar^{39} from Ti is four times its production in Fe. The Ar^{39} production in Ti has been measured only at 600 Mev, where it is three times that in Fe (R. DAVIS, private communication, 1972). From the fact that the mass of Ti is closer to the mass of Ar^{39} than is that of Fe, it is anticipated that the Ti cross-section will have a lower threshold than will the Fe cross section. Until the Ti cross-section is measured for protons of energies between 50 and 500 Mev, the simplest procedure is to make our assumption. Table 3 gives the results of this calculation for depths of ~1, ~5, ~15, and ~30 g/cm^2.

The Ar^{39} in Fe at all depths is 20 dpm/kg Fe, which is twice the value expected from galactic cosmic rays (REEDY and ARNOLD, 1972). If the excess Ar^{39} in Fe is attributed to solar flares, then there have been a surprisingly large number of very energetic (>200 Mev) protons in solar flares during the past 1000 years. The production cross section of Ar^{39} in Fe is negligible for protons <200 Mev (DAVIS *et al.*, 1963). The intensity of such energetic solar-flare protons equals that of galactic cosmic-ray nucleons (>200 Mev/nucleon). This result is quite similar to an estimate of the flux of solar-flare protons for the 19th solar cycle by MORE and TIFFANY (1962). The 20th solar cycle had fewer very energetic solar flares. The Ar^{39} activity in Fe indicates that the 19th solar cycle was more typical of solar activity during the past 1000 years than was the 20th.

The Ar^{39} in Ca decreased with increasing depth at less than 15 g/cm^2 and then increased at a depth of 30 g/cm^2, in a manner very similar to that observed for the Ar^{37}/Ca ratios in Apollo 12 and 14 samples. It appears that Ar^{37} and Ar^{39} in Ca arise from protons of similar energy. From the depth dependence of Ar^{39} in Ca, we estimate that 45 ± 10 dpm of Ar^{39}/kg Ca is produced at 1-g/cm^2 depth. STOENNER *et al.* (1971) measured the Ar^{39} production cross-section in Ca for 600-Mev protons to be 2 mb. If the Ar^{39} cross-section in Ca had a similar energy dependence to that of the Ar^{37} in Ca, then its cross-section would rise linearly from 0 for protons of approximately 45-Mev energy to 2 mb for 80-Mev energy protons and would be constant for higher energy protons. With these cross-section values, we calculate a solar-flare ($\geqq 50$ Mev) proton flux of 8×10^7/cm^2 year. Averaged over the past 1000 years, the intensity of solar flares responsible for Ar^{39} in Ca is then equivalent to three flares per year of the intensity of the flare of January 24, 1971.

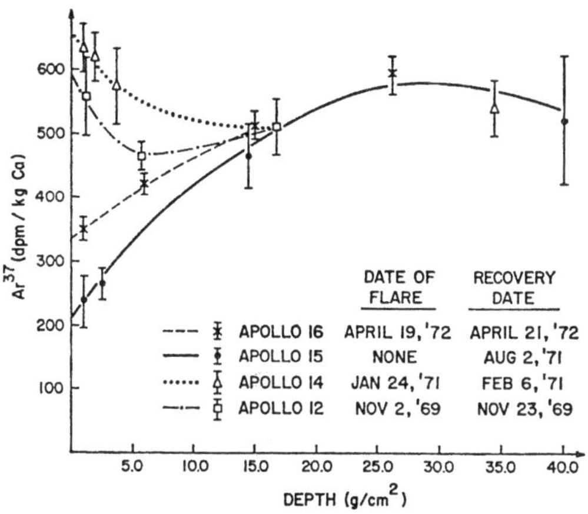

Fig. 3. The Ar^{37}/Ca ratio as a function of depth for material from the Apollo 12, 14, 15, and 16 missions

The Ar^{39} in K increases continuously with depth to our maximum depth of 30 g/cm^2. The nuclear reaction responsible for the Ar^{39} in K is probably $K^{39}(n, p)Ar^{39}$. The depth dependence of Ar^{39} in K indicates that the neutrons involved in this reaction have been moderated to such an extent that it will be necessary to solve the neutron-diffusion equation in lunar material before the solar-flare effect on K can be separated from the effect of galactic cosmic rays.

References

Begemann, F., Vilcsek, E., Rieder, R., Born, W., Wanke, H.: Proc. Apollo 11 Lunar Sci. Conf., Geochim. Cosmochim. Acta, Suppl. 1, **2**, 995 (1970).

Bostrom, C.: Solar-geophysical data, Washington, D. C.: U.S. Dept. of Commerce, 1971.

D'Amico, J., de Felice, J., Fireman, E. L., Jones, C., Spannagel, G.: Proc. Second Lunar Sci. Conf., Geochim. Cosmochim. Acta, Suppl. 2, **2**, 1825–1839 (1971).

Davis, R., Jr., Stoenner, R. W., Schaeffer, O. A.: Radioactive dating, p. 335. Vienna: International Atomic Energy Agency, 1963.

Finkel, R. C., Arnold, J. R., Imamura, M., Reedy, R. C., Loosli, J.S., Evans, J. C., Delany, A. C., Shedlovsky, J. P.: Proc. Second Lunar Sci. Conf., Geochim. Cosmochim Acta, Suppl. 2, **2**, 1773 (1971).

Fireman, E. L., D'Amico, J., de Felice, J.: Science, **167**, 566 (1970).

Fireman, E. L., D'Amico, J., de Felice, J., Spannagel, G.: Proc. Third Lunar Sci. Conf., Geochim. Cosmochim. Acta, Suppl. 3, **2**, in press (1972).

Kelley, G. D., Eldredge, J. S., Schonfeld, E., Bell, P. R.: Proc. Second Lunar Sci. Conf., Geochim. Cosmochim. Acta, Suppl. 2, **2**, 1747–1757 (1971).

More, K., Tiffany, O. L.: Proc. Symp. Protection Against Radiation Hazards in Space, Oak Ridge Natl. Lab., TID-7652, p. 682, 1962.

Rancitelli, L. A., Perkins, R. W., Felix, W. D., Wogman, N. A.: Proc. Second Lunar Sci. Conf., Geochim. Cosmochim. Acta, Suppl. 2, **2**, 1757–1773 (1971).

Reedy, R. C., Arnold, J. R.: Journ. Geophys. Res., **77**, 537 (1972).

Stoenner, R. W., Lyman, W. J., Davis, R., Jr.: Proc. Second Lunar Sci. Conf., Geochim. Cosmochim. Acta, Suppl. 2, **2**, 1813 (1971).

Solar Activity and Precipitation

By J. Xanthakis
Research Center for Astronomy and Applied Mathematics, Academy of Athens, Greece

With 8 Figures

Abstract

As it is known the up todate researches for finding out the correlation between solar activity—expressed by the relative sunspot numbers—and of the annual precipitation level at various localities, has shown that this correlation is, in general, small and insignificant.

In this paper we have taken as index of the solar activity, instead of the relative sunspot numbers, the areas index I_a being defined by the relation:

$$I_a = \frac{1}{2}[\sqrt{A} + \sqrt{f}]$$

where A and f are respectively the areas of sunspots and faculae corrected for foreshortening.

The observations of the annual precipitation R indicate that the dispersion diagrams between R and the areas index I_a in different regions present an X-wise formation outcoming of the fact that for the extreme values of the areas index I_a ($I_a < 20$, $I_a > 40$) the values of the annual precipitation R are separated into two distinguished groups from which the one involves small values of R and the other high values.

For the neutralization, as much as possible, of the various atmospheric factors which affect rainfall quite strongly we used the following method:

1) For every observational station we have calculated the differences $R - R_0$, where R represents the observed annual precipitation and R_0 the smallest value of the same quantity which has been noted during the time interval observed in each place.

2) Were duly considered all the observational stations lying within an Earth's surface zone defined by the parallels of latitude:

$$\varphi_i \leqq \varphi \leqq \varphi_{i+10^\circ}, \qquad i = 40^\circ, 50^\circ, 60^\circ, 70^\circ.$$

3) For each considered zone we have calculated the mean zonal annual precipitation defined with the help of the relation:

$$\overline{R - R_0} = \frac{1}{N} \sum_{i=1}^{N} (R_i - R_0)$$

where N represents the number of the stations considered in each zone.

4) The comparison of the values of the areas index I_a with the corresponding mean zonal annual precipitation $R-R_0$ at different zones of the surface of the earth indicates that there exists a quite satisfactory correlation, positive or negative, the correlation coefficient varying between 0.60 and 0.87.

Thus in the entire polar zone $80° \geqq \varphi \geqq 70°$ the correlation coefficient is equal to $r_{I_a, \overline{R-R_0}} = +0.77$.

In the zone $70° \geqq \varphi \geqq 60°$ the correlation coefficient is equal to $r_{I_a, \overline{R-R_0}} = -0.71$.

In the zone $60° \geqq \varphi \geqq 50°$ the correlation is at the beginning negative and then positive. The correlation coefficient are corresponding by equal to

$$r_{I_a, \overline{R-R_0}} = -0.69 \quad (1890\text{–}1914),$$

$$r_{I_a, \overline{R-R_0}} = +0.79 \quad (1914\text{–}1965).$$

A quite similar correlation exists in the zone $50° \geqq \varphi \geqq 40°$.

Introduction

The problem of correlation of solar activity with the various phenomena of the troposphere, such as the air temperature, the atmospheric pressure, the precipitation, etc. has been a matter of interest, from time to time, to some investigators. With regard to precipitation a limited research, conducted in independent locations, or in small areas of the Earth's surface proved that the correlation between solar activity—expressed by the Zürich relative sunspot numbers—and the annual precipitation is, in general, small or insignificant with the exception of a few particular cases.

According to DEPIETRI (1960) the relation between the 11-year cycle of solar activity expressed by the Wolf numbers and the pluviometric cycle with a period of 10.8 years in Spain, shows a difference of almost constant phase with a mean retardation of 2.1 ± 0.3 years. A distinct influence of solar activity upon the amount of annual precipitation in Spain two years later is evident, but this influence amounts only to 4%. According to ABBOT (1952), Wolf sunspot numbers above 20 are found to be connected with high average percentage precipitation in the first half of the year and low percentage in the second half. Wolf numbers below 20 show the opposite relation. Marked variations in precipitation with the Zürich annual relative sunspot numbers are found by PAPPE *et al.* (1962) in Southern Italy. Furthermore it is found that the amplitude of these variations does also depend on the average altitude of the 5 categories of stations considered. JOHNSON *et al.* (1964) have computed the correlation coefficients between the relative sunspot numbers and the total precipitation for 18 North American stations for the period 1883 to 1938. Negative values were obtained for all continental regions and positive values for oceanic and major-lake regions. However, the relative sunspot numbers are of rather small importance as a weather determining agent and may be neutralized by terrestrial phenomena.

From the above discussion which refer to relatively small areas of the Earth's surface we can conclude that one can expect a positive or negative correlation between the annual relative sunspot numbers and precipitation; this correlation

Table 1. Mean annual values of the areas index I_a

Years	I_a^{ob}	Years	I_a^{ob}	Years	I_a^{ob}	Years	I^{ob}
1881		1901	5.4	1921	23.8	1941	30.8
2	39.0	2	10.6	2	18.1	2	24.5
3	38.6	3	24.9	3	11.2	3	20.5
4	39.0	4	32.1	4	20.3	4	14.9
5	33.8	5	43.8	5	35.3	5	26.7
6	21.8	6	38.0	6	43.0	6	44.7
7	15.4	7	38.8	7	39.8	7	52.6
8	12.4	8	36.9	8	43.8	8	46.4
9	10.1	9	31.6	9	43.0	9	48.6
1890	13.7	1910	23.7	1930	31.5	1950	38.4
1	30.7	1	14.7	1	22.4	1	35.4
2	46.0	2	10.6	2	16.4	2	23.4
3	43.6	3	6.2	3	12.9	3	16.1
4	39.6	4	16.8	4	14.9	4	8.8
5	39.5	5	32.9	5	29.1	5	25.9
6	30.4	6	34.6	6	42.1	6	47.9
7	28.3	7	43.6	7	52.4	7	51.5
8	24.6	8	38.4	8	50.8	8	51.0
9	14.5	9	37.0	9	44.1	9	52.4
1900	11.0	1920	29.9	1940	35.6	1960	45.3

however, is, as a rule, very weak and may be disturbed in certain areas by various local influences of other terrestrial phenomena. In our previous work on certain solar phenomena, such as the intensity of the coronal line 5303 Å in the frequency range 1000 MHz$\leq f \leq$10000 MHz (Xanthakis, 1969), the variation of the equatorial diameter of the radio corona and the intensity of cosmic rays (Xanthakis, 1970) we have used as an index of solar activity, instead of the annual relative sunspot numbers, the areas index I_a defined with the help of the relation:

$$I_a = \tfrac{1}{2}[\sqrt{A} + \sqrt{f}] \tag{1}$$

where A and f are the total areas of sunspots and faculae, respectively, corrected for foreshortening, which are given by the relevant publications of the Greenwich Observatory.

The annual values of this new index for the time interval between 1881 and 1960, are given in Table 1.

Our previous work proved that the use of the areas index I_a instead of the relative sunspot numbers provides better correlations, and in some cases the combination of the areas index I_a, with the flare index I_f and the number of proton flares N_{PF} provides additional interesting relationships.

We shall use the areas index I_a as an index of solar activity, in an investigation of possible correlations between solar activity and precipitation.

Before taking up the main subject it is useful to study the dispersion diagrams between the areas index I_a and the annual precipitation in certain areas of Western Europe, with latitude between 48° N and 70° N.

Fig. 1 shows that the values of the annual precipitation R, in comparison to the corresponding values of the areas index I_a are distributed along two strips,

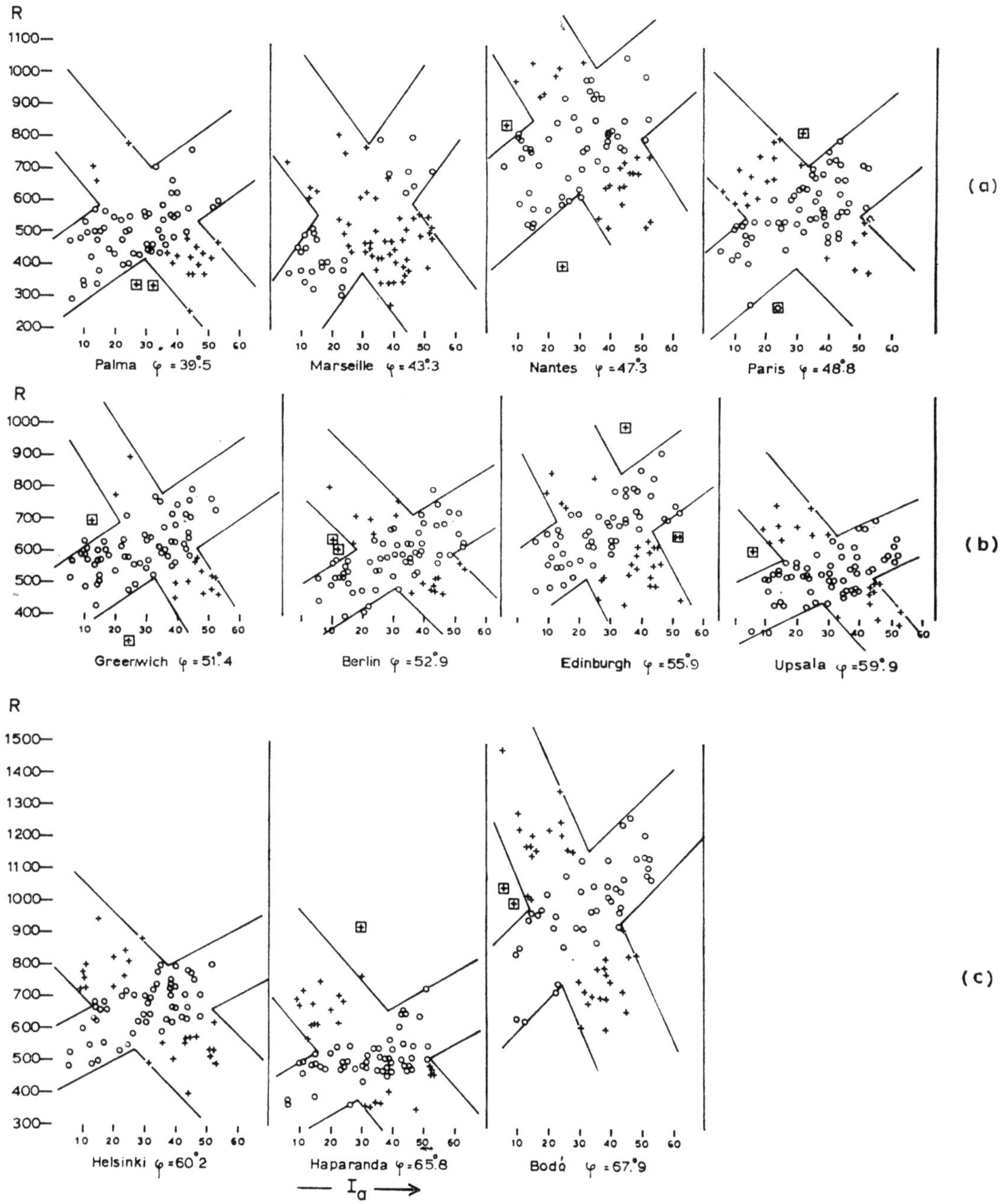

Fig. 1. Cross-wise formation of the dispersion diagrams between the annual precipitation R and the areas index I_a. The open circles represent the values of R offering a positive correlation with I_a and the crosses represent the values of R offering a negative correlation with I_a

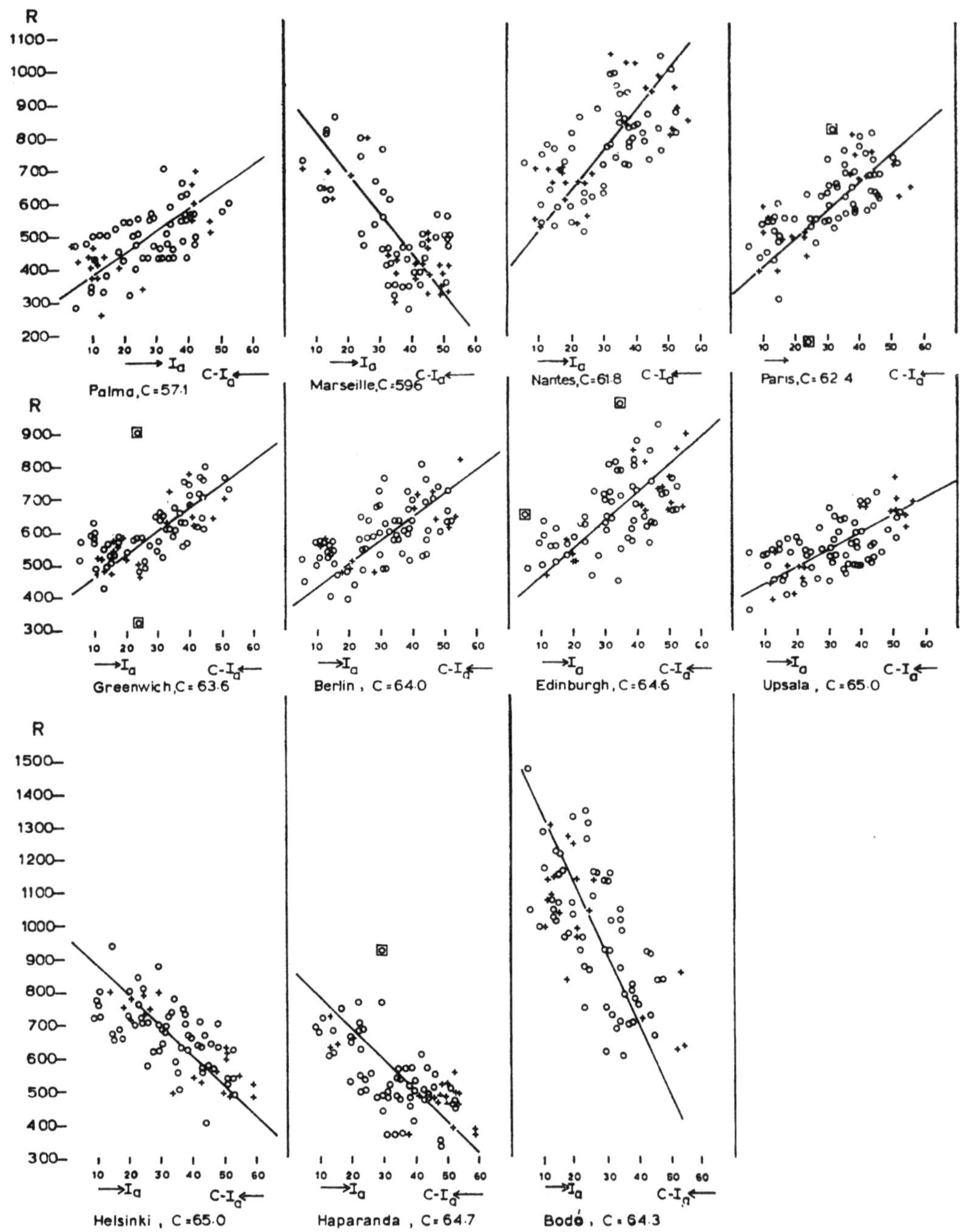

Fig. 2. Dispersion diagrams between the annual precipitation R and the areas index I_a as well as between R and the quantity $C-I_a$

crossing each other. A few values of R situated outside the two crossing strips are in small squares. This distribution indicates that the correlation between annual precipitation and solar activity is, for one and the same area, in some cases positive and in other cases negative. In other words for the extreme values of the areas index I_a i.e. $I_a<20$ and $I_a>40$ the corresponding values of R, are divided into two distinct groups, the one of which contains relatively small values of R, while the other contains high values.

From the cross-wise shape of the dispersion diagrams (R, I_a) it follows that the correlation coefficient between the two variables R and I_a, in each station separatelly assumes, in general, relatively small values, whether positive or negative.

This is, however, not the case, if we compare the values of the mean annual precipitation R to the areas index I_a, in the one hand, and on the other hand, to the quantity, $C-I_a$, where C is a constant quantity in each station.

This is shown in Fig. 2 where the small circles indicate the mean annual values of the precipitation R, these have a positive correlation with the corresponding values of the areas index I_a while the crosses indicate the values of R which have a positive correlation with the quantity $C-I_a$ i.e. a negative correlation with the quantity I_a. Next to the name of each station the numerical values of the constant C have been marked. From Fig. 2 we see that with the exception of Marseille (cf. Fig. 2a and b), the stations with latitude $\varphi \leqq 60°$ offering a positive correlation between R and I_a, are more numerous than those presented a negative correlation. In the three stations with latitude $\varphi > 60°$ (cf. Fig. 2c) the cases of negative correlation between R and I_a are, on the contrary, more numerous than those offering positive correlation.

It is also worth noticing that the constant C appears to vary according to the relation:

$$C = 65 \cos^2(\varphi - 60°). \tag{2}$$

Numerical values of constant C given in Fig. 2 have been computed with the help of relation (2).

Mean Zonal Annual Precipitation $\overline{R-R_0}$ Corresponding to Large Areas of the Earth's Surface

As it is known rainfall constitutes an extremely variable component, depending on many terrestrial factors, such as the general circulation of the atmosphere, the latitude and altitude of the station, the topography around the station and various other meteorological conditions. The contribution of these factors to the annual precipitation of a given area is very significant and eliminates or modifies the influence of the extraterrestrial factors, such as solar activity.

Consequently, if there should be a correlation between solar activity and precipitation, this correlation should be investigated not for each individual station, or within more or less limited areas but should extend to large zones of the Earth's surface, defined for example by two latitude parallels φ_1, φ_2. The influence of the terrestrial factors on the precipitation throughout such a zone is statistically greatly eliminated and thus the influence of extra-terrestrial factors is revealed.

We shall, therefore, use the following method for investigating correlations between solar activity and precipitation.

1) The meteorological stations have been divided into zones, defined by the relations:

$$\varphi_i \leqq \varphi \leqq \varphi_{i+10°}, \qquad i = 40°, 50°, 60°, 70°.$$

Only stations with more or less extensive series of precipitation data have been considered.

2) For each station the difference $R_i - R_0$ instead of the annual precipitation R, has been taken into consideration where R_0 is the smallest value of R recorded during the time interval covered by the precipitation observations for the station concerned.

3) For each year the sum $\sum_{i=1}^{N} (R_i - R_0)$ corresponding to the entire zone under study, as well as the mean value $\overline{R - R_0} = \frac{1}{N} \sum_{i=1}^{N} (R_i - R_0)$ where N represents the total number of the stations considered in each zone and for each year, have been computed.

It is obvious that the values of R_0 vary considerably from one station to another. In a station with high precipitation, for example, the value of R_0 is much greater than in another station of the same zone with low precipitation. The terrestrial factors however influencing R influence R_0 too in the same way. Therefore, the differences $R_i - R_0$ are to some extent free from local climatological influences. For the same reason the average $\overline{R - R_0}$ which shall call "Mean zonal annual precipitation in the zone (φ_i, $\varphi_{i+10°}$)" is also free to a great extent from the influence of terrestrial factors provided the number N of the stations considered is sufficiently great and these stations are uniformly distributed over the corresponding zone. In this way we conclude that if any influence of extraterrestrial factors on the precipitation should exists this influence should be mainly shown in the "mean zonal values of annual precipitation" $\overline{R - R_0}$ rather than in the values of R corresponding to each individual station which are greatly influenced by local factors.

In the following we shall proceed to study the variation of $\overline{R - R_0}$ in the following four zones of the Northern Hemisphere:

1) $70° \leqq \varphi \leqq 80°$,	Polar zone,	Long. 156° W to 80° E.
2) $60° \leqq \varphi \leqq 70°$,	Northern zone,	Long. 165° W to 40° E, 65° E to 177° E.
3) $50° \leqq \varphi \leqq 60°$,		Long. 176° W to 160° E.
4) $40° \leqq \varphi \leqq 50°$,		Long. 124° W to 152° E.

The values of annual precipitation R expressed in mm for all the stations studied have been taken from the volume "World Weather Records" (Smithsonian Miscellaneous collections vol. **79, 90** and **105**, for the period 1883–1940 and from the World Weather Records, vol. **1–6** published by the U.S. Department of Commerce, for the period 1940 onwards. The mean zonal values of the annual precipitation $\overline{R - R_0}$ have been smoothed with the help of the method $\frac{a+2b+c}{4}$.

I. Polar Zone $70° \leqq \varphi \leqq 80°$

Only very few stations with long and continuous series of precipitation observations are contained in this zone.

Table 2 contains the name, the geographical coordinates, the time interval of precipitation data and the value of R_0 for each station in this zone.

Out of the 12 stations in this zone, which are situated between the meridians 80° E to 156° W only 5 have continuous observations for more than 20 years. In most of them constant R_0 assumes values smaller or equal to 100.

Table 2a gives the values of $R-R_0$ during the interval 1912–1960 and for the individual stations as well as the "mean zonal values of annual precipitation" $\overline{R-R_0}$. In the last column the smoothed values of $\overline{R-R_0}$ are given.

The values of $R-R_0$ marked with an asterisc are not the observed values but the values smoothed out by using the formula: $(a+2b+c)/4$.

Fig. 3a illustrates the dispersion diagrams between $R-R_0$ and I_a in four stations of the Pólar zone separately, while Fig. 3b represents the dispersion diagram between the "mean zonal values of the annual precipitation" $\overline{R-R_0}$ in this zone and I_a. From these figures we perceive that the cross-wise shape revealed in the dispersion diagrams between $R-R_0$ and I_a of the three out of the four stations considered, does not appear any more in the dispersion diagrams $\overline{R-R_0}$, I_a, which shows a satisfactory positive correlation between these two variables.

The numerical results for this zone are shown in Fig. 4 where the continuous line illustrates the "mean zonal values of annual precipitation" $\overline{R-R_0}$ for the period 1912–1960 while the dashed line represents the annual values of the areas index I_a for the same time interval. The numbers under each year give the total number of stations considered for this year. We note that in spite of the limited number of the stations available in this zone, there exist a very satisfactory positive correlation between the quantities $\overline{R-R_0}$ and I_a.

The correlation coefficients between $\overline{R-R_0}$ and I_a on one hand and between $\overline{R-R_0}$ and the mean annual relative sunspot numbers (W.N.) on the other are

Table 2. Stations of the Polar Zone $70° \leqq \varphi \leqq 80°$

Name	Latitude	Longitude	Interval	R_0
Dickson	73° 31′ N	80° 23′ E	1922–1950	100
Vardo-Norway	70° 22′ N	31° 06′ E	1941–1960	400
Makkor fur-Norway	70° 42′ N	30° 25′ E	1951–1960	0
Ingoy-Norway	71° 04′ N	24° 09′ E	1941–1950	500
Isfjord Radio-Spitsbergen	78° 02′ N	14° 14′ E	1925–1960	100
Jan Mayen	71° 01′ N	8° 25′ E	1922–1950	300
Myggbukta-Greenland	73° 29′ N	21° 34′ E	1932–1940	100
Upernivik	72° 47′ N	56° 07′ E	1912–1932	100
Arctic Bay-Canada	73° 00′ N	85° 18′ E	1939–1960	0
Resolute Bay-Canada	74° 43′ N	94° 59′ E	1951–1960	0
Mould Bay-Canada	76° 14′ N	119° 20′ E	1951–1960	0
Barrow Alaska	71° 23′ N	156° 17′ E	1922–1960	0

Table 2a. Values of $\overline{R-R_0}$ in the Zone $80° \geqq \varphi \geqq 70°$ Long: 80° E to 160° W

Date	Upernivik	Dickson	Barrow	Isfjord Radio Spitsbergen	Jan Mayen	Myggbukta	Ingoy	Arctic Bay	Vardo	Resolute	Makkor fur	Mould Bay	N	$\overline{R-R_0}$	Smoothed values
1912	138												1	138	138
3	57												1	57	92
4	117												1	117	117
5	155														155
6	216														216
7	222														222
8	182														182
9	176												1	176	180
1920	186												1	186	181
1	177												1	177	165
2	219	76			68								3	121	130
3	154	39			113								3	102	101
4	80	50			110								3	80	99
5	128	60	216	126									4	133	117
6	116	62	67	244									4	122	133
7		38	58	354									3	150	152
8	185	43	67	419	207								5	184	170
9	139	141	103	288	135								5	161	155
1930	182	27	122		120								4	113	128
1		43*	74	114	270								4	125	115
2	96	74*	78	73*	276	−8							6	97	113
3		47	91	111*	287	128*							5	133	124
4		71	42	120*	250	180*							5	133	141
5			42	132*	347	140							4	165	153
6		79	48	251*	262	113							5	150	180
7		124	82	451	314	299							5	254	239
8		146	155	325	548	320							5	299	260
9			78	259	257	165		183					5	188	217
1940			75	373				157					3	202	187
1		148	102				87	202	237				4	155	161
2			165				234	178	46				4	133	134
3			137*		38			85	203				4	116	126
4		121	124*		180			136					4	140	130
5		156	75		149			109					4	122	138
6		76	112		303*			191	149*				5	166	166
7		128		176	430*		148	130	254				6	211	211
8		70	114	144	397*		353		113*				6	182	198
9		97	102	201	591		334	13	118*				7	208	180
1950		66	157	171	378		14	63	72*				7	132	152
1			137	218				97	11	123	275	76	7	134	128
2			102	196				113	117	98	85	71	7	112	120
3			72					141*	142	135	128	105	6	120	120
4			141	193				122*	119	132	91	96	7	128	135
5			117	208				111	150*	129	326	103	7	163	156
6			109	249				125	119*	137	127	43	7	169	173
7			185	214				158	287	146	260	74	7	190	175
8			184	327				109	83	97	177	71	7	150	170
9			171	442				144	260	145	133	77	7	196	175
1960			115	392				121	27	167	78	136	7	148	148

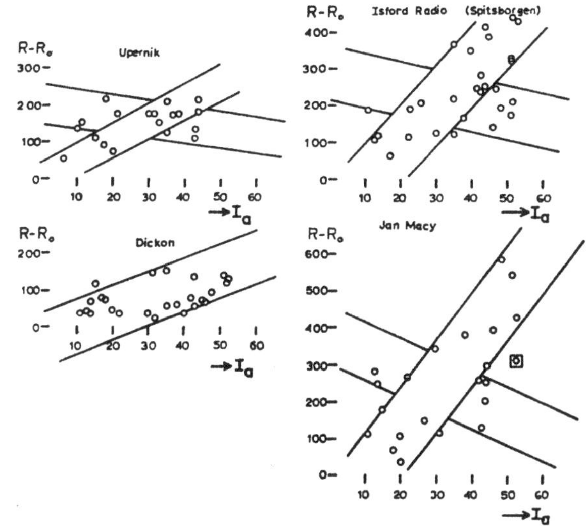

a

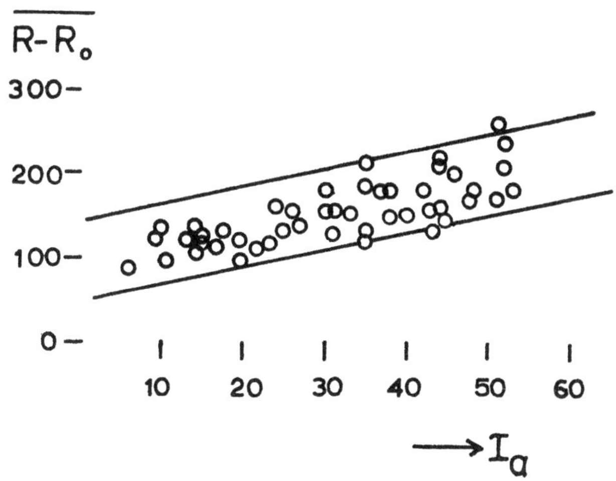

b

Fig. 3. Dispersion diagrams between $R-R_0$, I_a and between $\overline{R-R_0}$, I_a for four stations of the Polar Zone

respectively equal to

$$r_{\overline{R-R_0}, I_a} = +0.77$$

$$r_{\overline{R-R_0}, (\text{W.N.})} = +0.65$$

The above values of the correlation coefficients, however, should be considered as simple indications since the total number of stations is small ($N=12$) and the observations do not cover the entire period of 1912–1960.

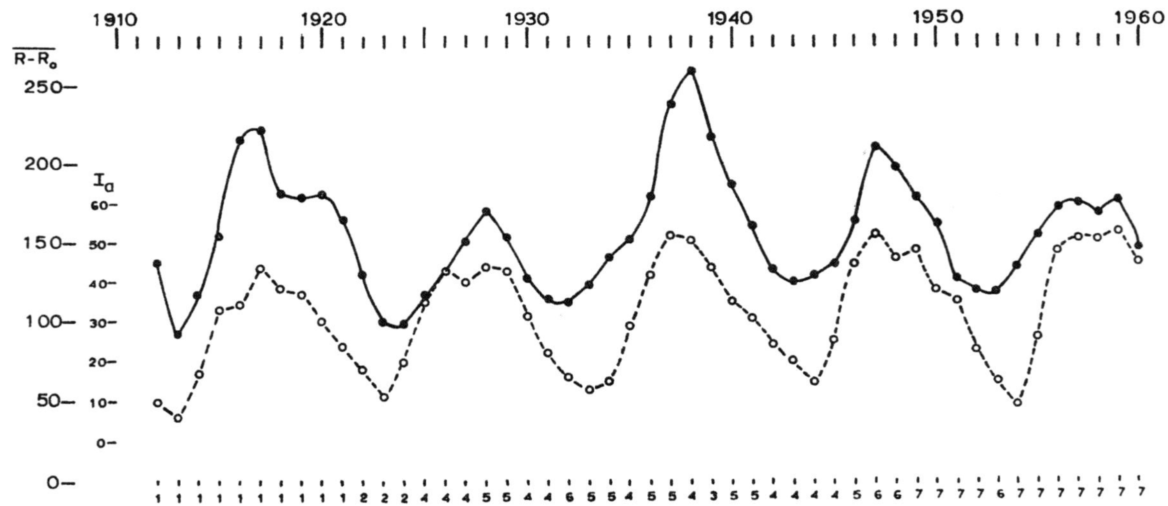

Fig. 4. Polar Zone: $70° \leqq \varphi \leqq 80°$. The continuous line represents the "mean zonal annual precipitation" $\overline{R-R_0}$ and the dashed line represents the variation of the mean annual values of the areas index I_a for the period 1912–1960

II. Northern Zone $60° \leqq \varphi \leqq 70°$, Long. 40° E to 165° W, 65° E to 177° E

More than half of the stations of this zone posses long series of precipitation observations starting with the year 1882 and only three for a time interval shorter than 20 years.

Table 3 gives the name, the geographical coordinates, the time-interval of precipitation observations and the value of R_0 for each station included between the meridian 40° E to 165° W in Northern Zone.

Table 3a gives for each year the values of the sum $\sum_{i=1}^{N}(R_i - R_0)$, the number N of the stations considered, the number n of the stations which were not taken into account when computing the sum $\sum_{i=1}^{N}(R_i - R_0)$ because some values of $R - R_0$ were considered irregular, and the "mean zonal values of annual precipitation" $\overline{R-R_0} = \frac{1}{N}\sum_{i=1}^{N}(R_i - R_0)$. In the last column the smoothed values of $\overline{R-R_0}$ are given.

For some of the stations included in the following Table 3b and for some years arround the maxima of solar activity, the corresponding values of $R - R_0$ are two or even more times greater than the corresponding values of $\overline{R-R_0}$. These values the number n of which is given in the fourth column of Table 3a were not taken into accound for the computation of the corresponding value of $\overline{R-R_0}$.

From the last column of Table 3a we perceive that the amplitude of the variation of $\overline{R-R_0}$ during the time-interval 1883–1960 is of the order of 100 mm.

Table 3. Stations of the Zone $70° \geqq \varphi \geqq 60°$, Long: 40° E to 165° W

Name	Latitude	Longitude	Intervalle	R_0
1. Archangelsk, U.S.S.R.	64° 35′ N	40° 36′ E	1882–1960	250
2. Helsinki, Finland	60° 10′ N	24° 57′ E	1882–1960	400
3. Östersund, Sweden	63° 11′ N	14° 39′ E	1882–1960	300
4. Haparanda, Sweden	65° 17′ N	24° 9′ E	1882–1960	350
5. Bodø, Norway	67° 17′ N	14° 24′ E	1882–1960	600
6. Karesuando	68° 27′ N	22° 30′ E	1882–1960	200
7. Berufjord, Ireland	64° 41′ N	14° 22′ W	1882–1960	800
8. Vestmanno, Ireland	63° 24′ N	20° 17′ W	1882–1960	1100
9. Stykkisholmur, Ireland	65° 05′ N	22° 46′ W	1882–1960	400
10. Angmagssalik Greenland	65° 37′ N	37° 33′ W	1882–1960	500
11. Ivigtut, Greenland	61° 12′ N	48° 15′ W	1882–1960	800
12. Jacobshavy, Greenland	69° 13′ N	51° 02′ W	1882–1960	130
13. Resolution Island	60° 35′ N	64° 53′ W	1931–1960	100
14. Clyde	70° 27′ N	68° 33′ W	1951–1960	100
15. Coppermine, Canada	67° 47′ N	115° 15′ W	1931–1960	200
16. Fort Good Hope, Canada	66° 15′ N	128° 38′ W	1918–1960	200
17. Dawson	64° 46′ N	139° 25′ W	1918–1960	200
18. Barter Island	70° 08′ N	143° 38′ W	1951–1960	100
19. Anchorage, Alaska	61° 13′ N	149° 51′ W	1942–1960	200
20. Tanana, Alaska	65° 10′ N	152° 06′ W	1918–1960	200
21. Bethel, Alaska	60° 47′ N	161° 41′ W	1931–1960	200
22. Nome, Alaska	64° 30′ N	165° 24′ W	1918–1960	250

Table 3 a

Years	$\Sigma(R-R_0)$	N	n	$\overline{R-R_0}$	Smoothed values	Years	$\Sigma(R-R_0)$	N	n	$\overline{R-R_0}$	Smoothed values
1883	2417	11		220		1922	4866	16		384	299
4	2544	11		232	218	3	4138	16		259	267
5	2097	11		191	203	4	3703	15		247	253
6	2203	11		200	212	5	3912	15		261	249
7	2309	9		256	236	6	3648	16		228	230
8	2566	11		233	260	7	3058	15		204	216
9	3507	11		319	294	8	3203	14		229	231
1890	3336	11		304	285	9	3950	15		263	263
1	2348	11		213	202	1930	4176	14		298	285
2	1911	11		174	202	1	4759	17		280	292
3	2679	11		243	217	2	5564	18		309	300
4	2302	11		209	227	3	4814	16		301	302
5	2456	10		246	243	4	4751	16		297	297
6	2970	11		270	258	5	4978	17		293	290
7	2718	11		247	280	6	4988	18		277	262
8	3917	11		356	317	7	3213	16	2	201	227
9	3085	10		308	309	8	3430	15	3	229	220
1900	2366	9		263	290	9	4153	19		219	219
1	3572	11		325	308	1940	3707	17		218	213
2	3512	11		319	308	1	3764	19		198	224
3	3198	12		267	255	2	5639	20		282	266
4	2920	12		244	242	3	5552	19		293	297
5	1661	9	3	185	204	4	6423	20		326	317
6	1610	8	4	201	198	5	6636	28		332	323
7	2466	12		206	218	6	5803	19		306	285
8	3127	12		261	245	7	3895	18		216	229
9	3024	12		252	243	8	2292	10	3	229	222
1910	2243	11		208	234	9	2978	14		216	219
1	2932	11		267	243	1950	2917	15		194	210
2	2774	12		231	257	1	3979	17		234	226
3	3277	11	1	298	298	2	4209	18		234	246
4	2719	12		227	235	3	4213	15		281	274
5	2260	12		188	202	4	3868	13		299	277
6	2458	12		205	196	5	3914	17		230	234
7	2231	12		186	192	6	3297	18		183	195
8	2694	14		192	187	7	3494	19		184	185
9	2526	14		180	195	8	3566	19		188	194
1920	3386	15		226	240	9	3857	18		214	205
1	4933	15		329	297	1960	3740	19		197	197

Remark: In certain Stations, instead of the observed values of the differences $R-R_0$ we have taken the corresponding values of $R-R_0$ in pairs smoothed out with the help of the formula $\frac{a+2b+C}{4}$.

Furthermore the variation of the values of $\overline{R-R_0}$ during the years 1883–1960 is opposite to the variation of the corresponding values of I_a. It is rendered evident that in the Northern zone the two quantities $\overline{R-R_0}$ and I_a are negatively correlated, while in the Polar Zone we have are opposite effect.

Table 3b. Irregular values of $R-R_0$ in some station of the Northern Zone

Station	Years					
	1905	1906	1913	1937	1938	1948
Archangelsk	507					501
Karesuanda			61			
Berufjord	471	504		794	500	
Vestmanno		485				458
Angmagsalik	399	401				
Ivigtut		496		717	679	
Bodó					669	669
$\overline{R-R_0}$	229	201	298	201	229	222
N	9	8	11	16	15	10
$\sigma\pm$	75	87	113	109	92	83

Table 3c

Stations	Latitude	Longitude	Interval	R_0
Berezov	63° 56′ N	65° 04′ E	1926–1960	300
Khanty-Mansiyak	60° 58′ N	69° 04′ E	1951–1960	400
Surgut	61° 15′ N	73° 24′ E	1926–1960	300
Turukhansk	65° 47′ N	87° 57′ E	1929–1960	350
Olekminsk	60° 22′ N	120° 26′ E	1926–1960	150
Podkamennay-Tunguska	61° 36′ N	90° 00′ E	1951–1960	350
Tura	64° 10′ N	100° 04′ E	1951–1960	250
Olenek	68° 30′ N	112° 20′ E	1951–1960	150
Shelagontsy	66° 15′ N	114° 17′ E	1951–1960	200
Vilyuysk	63° 46′ N	121° 37′ E	1951–1960	150
Yakutsk	62° 22′ N	129° 43′ E	1926–1960	100
Verkhoyansk	67° 35′ N	133° 50′ E	1926–1960	70
Ust Mayskoe	60° 23′ N	134° 27′ E	1926–1960	150
Oymyakon	63° 18′ N	143° 09′ E	1951–1960	150
Zyryanka	65° 44′ N	150° 54′ E	1951–1960	200
Seymchan	62° 55′ N	152° 25′ E	1951–1960	200
Ilirney	67° 20′ N	168° 14′ E	1951–1960	150
Ust Apuka	60° 27′ N	169° 35′ E	1951–1960	400
Markovo	64° 41′ N	170° 25′ E	1951–1960	200
Novo-Mariinsky Post	64° 47′ N	177° 34′ E	1926–1960	100

In addition to the 22 stations included in Table 3, we also took into consideration 20 stations of Northern Zone located in Northern Asia (U.S.S.R.) between the meridians of 65° E to 177° E. The names and other data related to these stations are given in Table 3c. The values of the sum $\sum_{i=1}^{N}(R_i-R_0)$ the total number N of the stations considered and the corresponding values of $\overline{R-R_0}$, are given in Table 3d.

The numerical results for this zone are shown in Fig. 5a and b.

Table 3d

Years	$\sum_{i=1}^{N}(R_i-R_0)$	N	$\overline{R-R_0}$	Smoothed values	Years	$\sum_{i=1}^{N}(R_i-R_0)$	N	$\overline{R-R_0}$	Smoothed values
1926	715	6	119	119	1944	1035	8	134	120
7	530	6	88	105	5	819	8	102	110
8	884	7	126	114	6	860	8	108	102
9	805	7	115	107	7	726	8	91	102
1930	564	8	71	82	8	835	7	119	98
1	427	6	71	72	9	448	7	64	98
2	378	5	76	84	1950	997	7	142	117
3	456	4	114	98	1	1640	18	91	91
4	441	5	88	102	2	1750	19	92	97
5	1086	7	155	130	3	2220	20	111	108
6	732	7	105	115	4	2105	18	117	113
7	615	7	88	102	5	2052	19	108	109
8	759	6	126	113	6	2040	20	102	109
9	811	7	116	116	7	2460	20	123	120
1940	759	7	108	102	8	2620	20	131	122
1	692	8	87	95	9	2100	20	105	111
2	817	8	102	101	1960	2860	19	124	124
3	928	8	116	117					

In Fig. 5a the continuous line represents the values of the "mean zonal values of annual precipitation" $\overline{R-R_0}$, corresponding to the 22 stations of the Northern Zone contained between 40° E to 165° W while the dashed line represents the corresponding values of the areas index I_a. The crosses placed in the years 1905, 1906, 1937 and 1948 offer the values of the quantity $\overline{R-R_0}$ found if the irregular values of $R-R_0$ given in the Table 3b, are also taken into account.

In Fig. 5b the continuous line illustrates the values of the quantity $\overline{R-R_0}$ corresponding to the 20 stations of the Northern Zone contained between the meridians 65° E to 177° E (Northern Asiatic Continent).

From the data given in Table 3a the following values of the correlation coefficients between $\overline{R-R_0}$, and I_a on one hand, and between $\overline{R-R_0}$ and the Wolf numbers (W.N.) on the other have been found.

$$r_{\overline{R-R_0},\,I_a} = -0.71$$

$$r_{\overline{R-R_0},\,(\text{W.N.})} = -0.66$$

If we would also take into account, for the computation of the differences $\overline{R-R_0}$, the irregular values given in Table 3b. These correlation coefficients assume the values

$$r_{\overline{R-R_0},\,I_a} = -0.64$$

$$r_{\overline{R-R_0},\,(\text{W.N.})} = -0.63$$

In Fig. 5b the values of the quantity $\overline{R-R_0}$ show slight fluctuations around a mean value without any noticeable correlation with the areas index I_a.

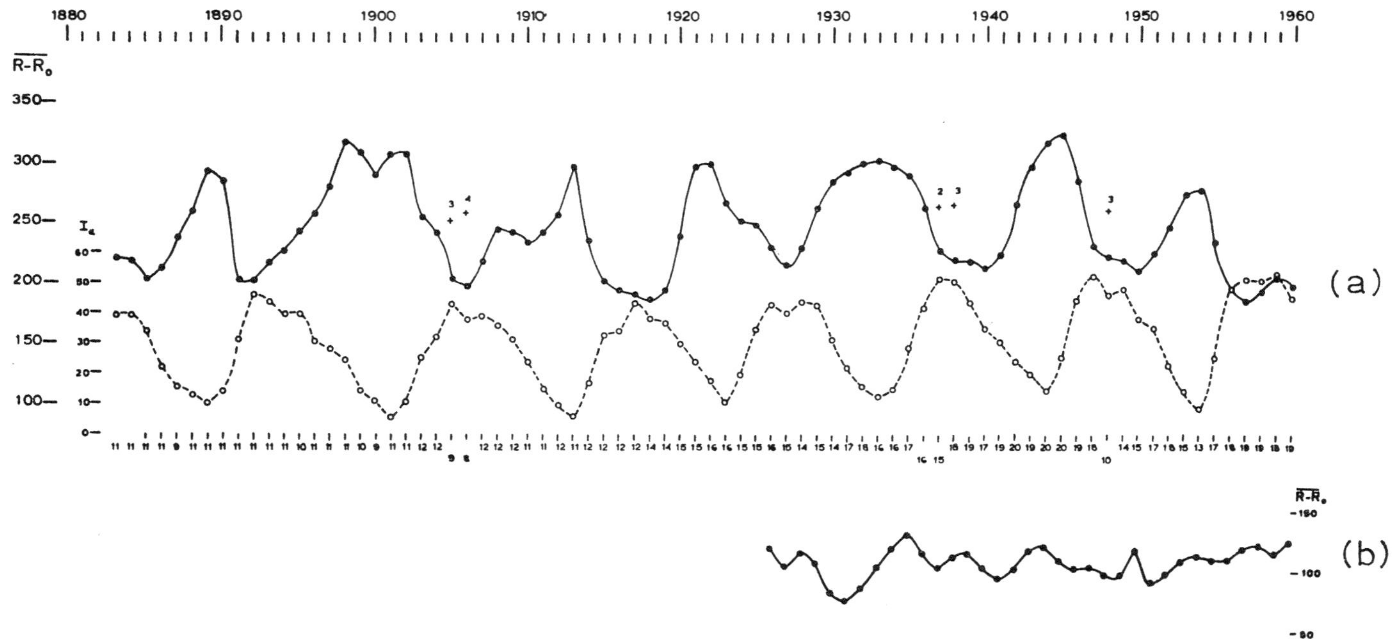

Fig. 5a and b. Northern Zone: $70° \leqq \varphi \leqq 80°$, long. 40° E to 165° W, 65° E to 177° E. a (Upper part). The continuous line represents the values of the "mean zonal annual precipitation" $\overline{R-R_0}$ corresponding to 22 stations of the Northern Zone included between 40° E to 165° W and the dashed line represents the mean annual variation of the areas index I_a. The crosses placed above the years 1905, 1937 and 1948 represent the values of $\overline{R-R_0}$ when some irregular values of $R-R_0$ were also used in the computation. b (Lower part). The continuous line represents the values of $\overline{R-R_0}$ corresponding to 20 stations of the Northern Zone situated between 65° E to 177° E (Northern Asiatic Continent)

III. Zone 50° ≦ φ ≦ 60°, Long. 176° W to 160° E

The name and the other data concerning the stations included in this zone are given in Table 4.

This zone has a much larger area than the previous ones. Moreover for most of the stations in this zone we do posses long series of precipitation observations. Furthermore these stations are more or less uniformly distributed throughout the entire zone.

Table 4a gives for each year of the time interval 1890 to 1960 the values of the sum $\sum_{i=1}^{N}(R_i-R_0)$, the corresponding number N of stations considered, the number n of the irregular values of $R-R_0$, the "mean zonal values of annual precipitation" $\overline{R-R_0}$ and the smoothed values of $\overline{R-R_0}$.

Likewise in some stations of this zone and particularly during the years of minimum solar activity the values of the differences $R-R_0$ differ significantly from the corresponding "mean zonal values of annual precipitation" $\overline{R-R_0}$. These irregular values are presented in Table 4b, and number n for each corresponding year is given in the fourth column of Table 4a. From Table 4b for example we see that during the years of minimum solar activity 1922–1923, 1943–1944, 1953–1954, the values of $R-R_0$ corresponding to some stations are two to seven times greater than the corresponding value of $\overline{R-R_0}$ as computed from the remaining stations (Table 4a).

The results derived for this zone are shown in Fig. 6a, b.

Table 4. Stations of the Zone 60° ≧ φ ≧ 50° Long: 116° W to 160° E

Name	Latitude	Longitude	Interval	R_0
1. Adak, Alaska	51° 53′ N	176° 39′ W	1943–1960	900
2. St. Paul Island, Alaska	57° 15′ N	170° 10′ W	1918–1960	400
3. Kodiak Alaska	57° 45′ N	152° 22′ W	1903–1960	1200
4. Dutch Harbor, Alaska	53° 55′ N	166° 30′ W	1908–1954	1200
5. Cold Bay	55° 12′ N	162° 43′ W	1943–1960	500
6. King Salmon	58° 41′ N	156° 39′ W	1943–1960	200
7. Yakutat, Alaska	59° 31′ N	139° 40′ W	1941–1960	2200
8. Sitka, Alaska	57° 04′ N	135° 19′ W	1900–1960	1900
9. Juneau, Alaska	58° 18′ N	134° 24′ W	1912–1960	1200
10. Annette, Alaska	55° 02′ N	131° 34′ W	1942–1960	2000
11. Langara, Canada	54° 15′ N	133° 03′ W	1941–1960	1200
12. Massett, Canada	53° 58′ N	132° 09′ W	1900–1952	700
13. Sandspit, Canada	53° 15′ N	131° 49′ W	1949–1960	1000
14. Prince Rupert, Canada	54° 17′ N	130° 23′ W	1941–1960	1800
15. Port Hardy, Canada	50° 41′ N	127° 22′ W	1945–1960	1200
16. Bella Coola, Canada	52° 40′ N	126° 54′ W	1901–1960	900
17. Prince George, Canada	53° 53′ N	122° 41′ W	1941–1960	400
18. Kamloops, Canada	50° 41′ N	120° 29′ W	1895–1960	150
19. Barkerville, Canada	53° 02′ N	121° 35′ W	1890–1960	700
20. Calgary, Canada	51° 02′ N	114° 02′ W	1890–1960	200
21. Edmonton, Canada	53° 33′ N	113° 30′ W	1890–1960	200

Table 4 (continued)

Name	Latitude	Longitude	Interval	R_0
22. Fort McMurray, Canada	56° 39′ N	111° 13′ W	1942–1960	200
23. Prince Albert, Canada	53° 10′ N	105° 38′ W	1890–1960	200
24. Qu'Appelle, Canada	50° 30′ N	103° 47′ W	1890–1960	250
25. Trout Lake, Canada	53° 50′ N	89° 52′ W	1941–1960	400
26. Armstrong, Canada	50° 17′ N	88° 54′ W	1943–1960	500
27. Sept Iles, Canada	50° 13′ N	66° 16′ W	1945–1960	800
28. Moosonee, Canada	51° 31′ N	80° 31′ W	1941–1960	500
29. Nitcheguon, Canada	53° 12′ N	70° 54′ W	1943–1960	500
30. Fort Chimo, Canada	58° 06′ N	68° 26′ W	1951–1960	100
31. Goose Bay, Canada	53° 19′ N	60° 25′ W	1942–1960	600
32. Port Harrison, Canada	58° 25′ N	78° 21′ W	1941–1960	200
33. Knob, Lake, Canada	54° 48′ N	66° 49′ W	1951–1960	500
34. Hopedale, Canada	55° 27′ N	60° 14′ W	1949–1960	500
35. Belle Isle, Canada	51° 53′ N	55° 22′ W	1951–1960	350
36. Churchill, Canada	58° 46′ N	94° 13′ W	1941–1960	200
37. Valentia, Ireland	51° 56′ N	10° 15′ W	1890–1960	1200
38. Edinburgh, U.K.	55° 55′ N	3° 11′ W	1890–1960	450
39. Aberdeen, U.K.	57° 10′ N	2° 06′ W	1890–1960	600
40. Greenwich, U.K.	51° 29′ N	0° 00′	1890–1960	400
41. Frankfurt, Germany	50° 07′ N	8° 39′ E	1890–1960	400
42. Oslo, Norway	59° 56′ N	10° 44′ E	1890–1960	450
43. Copenhagen, Denmark	55° 41′ N	12° 38′ E	1890–1960	400
44. Berlin, Germany	52° 27′ N	13° 18′ E	1890–1960	400
45. Breslau	51° 07′ N	17° 05′ E	1890–1960	350
46. Uppsala, Sweden	59° 48′ N	17° 38′ E	1890–1960	400
47. Königsberg	54° 44′ N	20° 34′ E	1890–1944	500
48. Warszawa, Poland	52° 13′ N	21° 03′ E	1894–1960	350
49. Vilnjus (Wilno)	54° 41′ N	25° 15′ E	1890–1960	350
50. Leningrad, U.S.S.R.	59° 58′ N	30° 18′ E	1890–1960	400
51. Kiev, U.S.S.R.	50° 20′ N	30° 30′ E	1890–1960	400
52. Moskva, U.S.S.R.	55° 50′ N	37° 33′ E	1890–1960	350
53. Nikolaewkoe-Oktjabrsky Gorodok, U.S.S.R.	51° 27′ N	45° 27′ E	1890–1960	250
54. Kasan, U.S.S.R.	55° 47′ N	49° 08′ E	1890–1960	300
55. Orenbourg, U.S.S.R.	51° 45′ N	55° 06′ E	1890–1960	220
56. Perm (Molostov) U.S.S.R.	58° 01′ N	56° 16′ E	1890–1960	400
57. Swerdlowsk-Ekaterinburg, U.S.S.R.	56° 50′ N	60° 38′ E	1890–1960	250
58. Tobolsk, U.S.S.R.	58° 12′ N	68° 14′ E	1921–1960	250
59. Akmolinsk, U.S.S.R.	51° 12′ N	71° 23′ E	1923–1960	150
60. Omsk, U.S.S.R.	55° 01′ N	73° 19′ E	1921–1960	200
61. Barnaul, U.S.S.R.	53° 20′ N	83° 48′ E	1921–1960	300
62. Tomsk, U.S.S.R.	56° 30′ N	84° 58′ E	1921–1960	350
63. Minusinsk, U.S.S.R.	53° 45′ N	91° 41′ E	1921–1960	200
64. Yenisseysk, U.S.S.R.	58° 27′ N	92° 10′ E	1921–1960	350
65. Kirensk, U.S.S.R.	57° 47′ N	108° 07′ E	1922–1960	250
66. Tchita, U.S.S.R.	59° 02′ N	113° 30′ E	1924–1960	200
67. Blagovyeshtchensk, U.S.S.R.	50° 17′ N	127° 30′ E	1921–1960	300
68. Nikolayevsk on Amur, U.S.S.R.	53° 08′ N	140° 45′ E	1926–1960	300
69. Okhotsk, U.S.S.R.	59° 21′ N	143° 17′ E	1927–1960	150
70. Irkutsk, U.S.S.R.	52° 16′ N	104° 19′ E	1921–1960	250
71. Petropavlovsk, U.S.S.R.	52° 53′ N	158° 42′ E	1927–1960	400

Table 4a

Years	$\sum_{i=1}^{N}(R_i-R_0)$	N	n	$\overline{R-R_0}$	Smoothed	Years	$\sum_{i=1}^{N}(R_i-R_0)$	N	n	$\overline{R-R_0}$	Smoothed
1890	4968	25		199	199	1926	11258	43		262	267
1	4089	25		164	168	7	13315	46		289	275
2	3427	24		143	154	8	11196	43		260	255
3	4194	25		168	164	9	8871	42		211	237
4	4637	26		178	174	1930	10689	40		267	241
5	4591	27		170	167	1	10052	46		219	234
6	4030	27		149	155	2	10255	44		233	214
7	4115	27		152	155	3	7699	45		171	196
8	4504	27		167	178	4	9428	45		210	207
9	6082	27		225	211	5	11151	47		237	235
1900	6620	29		228	233	6	12072	47		257	261
1	5275	21	7	251	232	7	11768	40		294	285
2	5968	30		199	229	8	11198	38		295	296
3	7983	30		266	220	9	11488	38		302	284
4	4358	29		150	190	1940	10206	43		237	253
5	5488	28		196	186	1	11313	48		236	237
6	5819	29		201	192	2	12789	53		241	223
7	5127	30		171	192	3	8634	50	4	173	198
8	7190	32		225	199	4	9796	48	5	204	213
9	5559	32		174	191	5	15587	57		273	259
1910	5732	30		191	179	6	17233	60		287	299
1	4991	31		161	188	7	20545	59		348	325
2	7700	32		241	223	8	18311	58		316	320
3	7462	30		249	249	9	18381	61		301	294
4	8447	33		256	260	1950	15752	61		258	262
5	8401	30		280	275	1	14302	62		231	257
6	8768	31		283	278	2	19519	63		310	268
7	7968	30		266	270	3	12144	55	8	221	236
8	8797	33		267	262	4	10497	54	6	194	228
9	8008	32		250	237	5	18171	60		303	279
1920	5677	31		183	202	6	19511	62		315	305
1	8095	42		193	184	7	17742	62		286	303
2	5662	34	7	167	178	8	20096	62		324	320
3	6745	36	4	187	195	9	21222	61		348	342
4	10014	42		238	229	1960	21582	62		348	348
5	10928	43		254	252						

In Fig. 6a the continuous line represents the values of the quantity $\overline{R-R_0}$ while the dashed line gives the corresponding values of the areas index I_a. The crosses placed above the years 1901, 1922, 1923, 1943, 1944, 1953, and 1954, illustrate the values found for $\overline{R-R_0}$ if the irregular values of the $R-R_0$ given in Table 4b are also taken into account.

From Fig. 6a we see that the "mean zonal values of annual precipitation" $\overline{R-R_0}$ in the zone considered present a negative correlation with the area index I_a during the interval 1890–1913 and a positive correlation during the remaining interval

Table 4b. Irregular Values of $R-R_0$ in the Zone $50° \leqq \varphi \leqq 60°$

Stations	Values of $R-R_0$						
	1901	1922	1923	1943	1944	1953	1954
Kamloops	30						
Greenwich	115						
Aberdeen	108						
Uppsala	−3						
Leningrad	3						
Berlin	114	316					
Kasan	47						
Frankfurt		490	367		302		
Kiev		390	448				
Massett		585	843			1464	
Juneau		880		1802			
Bella Coola		613				828	952
Nikolaewskoe		431	430				
Swerdlowsk				320			
Nikolayevsk on Amur				393	377		
Vilna				323			
Adak						1182	1445
Oslo					403		
Yakutat				2088	1140		
Annete					664	938	829
Prince Rupert						1028	1152
Port Hardy						880	663
Bella Isle						688	755
						673	
$\overline{R-R_0}$	251	190	187	172	206	220	199
N	21	34	36	50	48	56	55

1914–1960. The Fig. 6b shows the "zonal values of annual precipitation" $\overline{R-R_0}$ for the Northern American Continent (Canada, Alaska) and of the corresponding values of the areas index I_a. It is worth mentioning that the reversal of the correlation between solar activity and precipitation is far more obvious in this area.

The correlation coefficient between $\overline{R-R_0}$ and I_a, and between $\overline{R-R_0}$ and the wolf numbers (W. N.) are respectively:

$$r_{\overline{R-R_0}, I_a} = -0.76 \quad \text{Period 1890–1913}$$

$$r_{\overline{R-R_0}, (\text{W.N.})} = -0.68 \quad \text{Period 1890–1913}$$

$$r_{\overline{R-R_0}, I_a} = +0.81 \quad \text{Period 1914–1960}$$

$$r_{\overline{R-R_0}, (\text{W.N.})} = +0.86 \quad \text{Period 1914–1960}$$

If for the computation of the "mean zonal values of annual precipitation" $\overline{R-R_0}$ the irregular values of the $R-R_0$ given in the Table 4b are also taken into

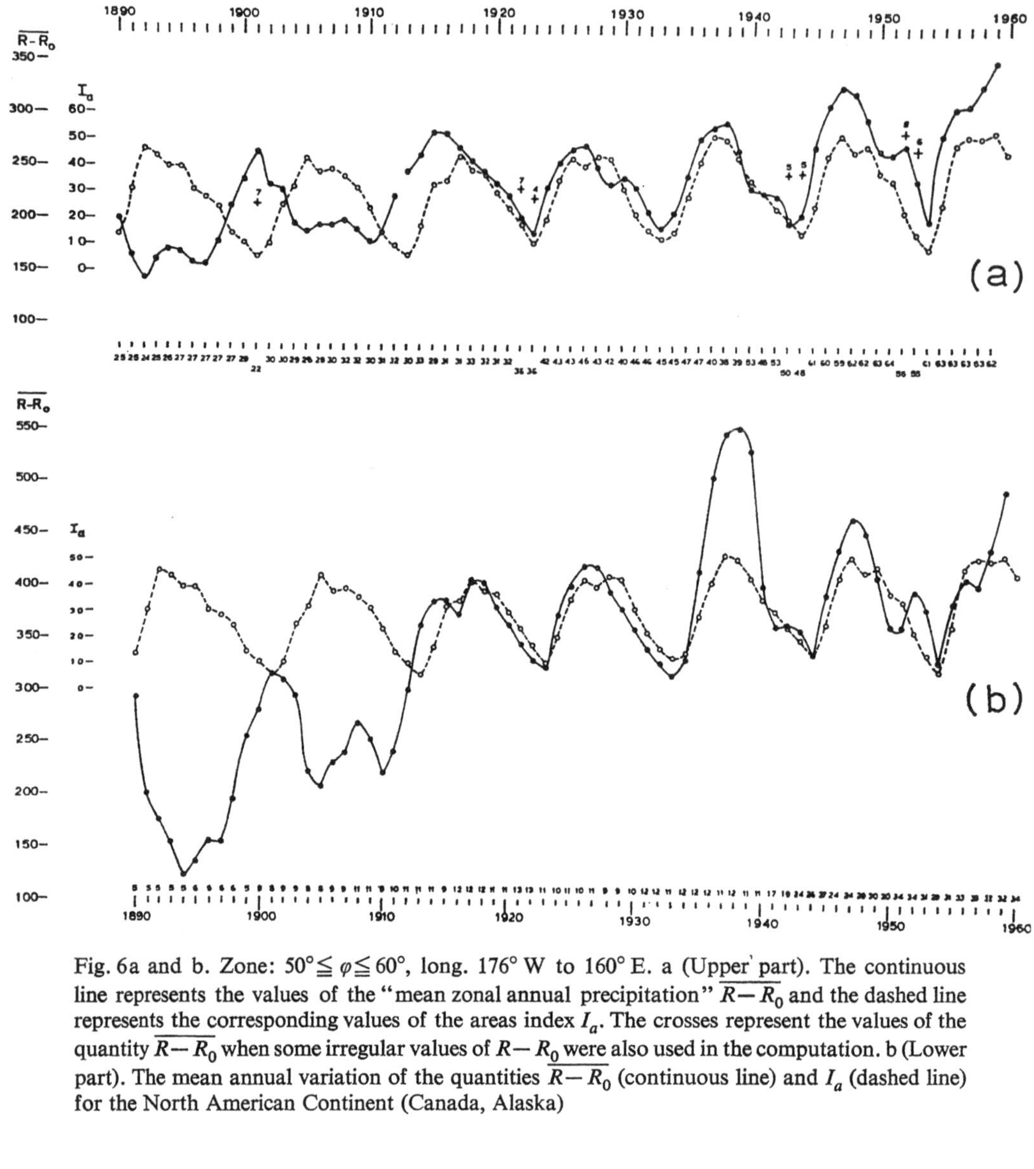

Fig. 6a and b. Zone: $50° \leqq \varphi \leqq 60°$, long. 176° W to 160° E. a (Upper part). The continuous line represents the values of the "mean zonal annual precipitation" $\overline{R-R_0}$ and the dashed line represents the corresponding values of the areas index I_a. The crosses represent the values of the quantity $\overline{R-R_0}$ when some irregular values of $R-R_0$ were also used in the computation. b (Lower part). The mean annual variation of the quantities $\overline{R-R_0}$ (continuous line) and I_a (dashed line) for the North American Continent (Canada, Alaska)

account, then the correlation coefficient slightly decrease for the period 1890–1913 and more so for the period 1914–1960 (positive correlation).

$$r_{\overline{R-R_0}, I_a} = -0.72 \quad \text{Period 1890–1913}$$

$$r_{\overline{R-R_0}, (\text{W.N.})} = -0.67 \quad \text{Period 1890–1913}$$

$$r_{\overline{R-R_0}, I_a} = +0.63 \quad \text{Period 1914–1960}$$

$$r_{\overline{R-R_0}, (\text{W.N.})} = +0.74 \quad \text{Period 1914–1960}$$

IV. Middle Latitude Zone $40° \leqq \varphi \leqq 50°$, Long. 124° W to 162° E

As in the previous zones, in the case of the middle Latitude Zone the majority of the stations studied have precipitation observations covering a rather long period of time and are uniformly distributed throughout the zone, which extends between 124° W to 162° E.

The name and the other data concerning the stations included in this zone are given in Table 5.

Table 5a gives for each year the values of $\sum_{i=1}^{N}(R_i - R_0)$, the number N of stations considered, the number n of the irregular values of $R - R_0$, the mean zonal values of annual precipitation $\overline{R - R_0}$ and the smoothed values of $\overline{\overline{R - R_0}}$.

In computing the "mean zonal values of annual precipitation" $\overline{R - R_0}$ given in the Table 5, the "irregular" values of $R - R_0$ were not taken into account. These irregular values are given in the Table 5b.

Table 5. Name and Data of the Stations in the Zone $50° \geqq \varphi \geqq 40°$ Long: 123° W to 151° E

Name	Latitude	Longitude	Intervalle	R_0
1. Victoria B. C., Canada	48° 24′ N	123° 19′ W	1886–1960	400
2. Portland Oreg., U.S.A.	45° 32′ N	122° 41′ W	1886–1960	700
3. Red Bluff, U.S.A.	40° 10′ N	122° 15′ W	1886–1960	300
4. Spokane, U.S.A.	47° 40′ N	117° 25′ W	1886–1960	250
5. Helena, U.S.A.	46° 34′ N	112° 04′ W	1886–1960	150
6. Salt Lake City, U.S.A.	40° 16′ N	111° 54′ W	1886–1960	250
7. Cheyenne, U.S.A.	41° 08′ N	104° 48′ W	1886–1960	250
8. North Platte, U.S.A.	41° 08′ N	100° 45′ W	1886–1960	250
9. Bismarck, U.S.A.	46° 47′ N	100° 38′ W	1886–1960	200
10. Winnipeg, Canada	49° 53′ N	97° 07′ W	1886–1960	300
11. Omaha, U.S.A.	41° 16′ N	95° 56′ W	1886–1960	450
12. Oregon Mo, U.S.A.	44° 58′ N	95° 11′ W	1886–1960	550
13. St. Paul, U.S.A.	44° 58′ N	93° 03′ W	1886–1960	400
14. Ouluth, U.S.A.	46° 47′ N	92° 06′ W	1886–1960	450
15. Peoria, U.S.A.	40° 43′ N	89° 36′ W	1886–1960	600
16. Chicago, U.S.A.	41° 53′ N	87° 37′ W	1886–1960	600
17. Marquette, U.S.A.	46° 34′ N	87° 24′ W	1886–1960	600
18. Detroit, U.S.A.	42° 20′ N	83° 03′ W	1886–1950	600
19. Toronto, Canada	43° 40′ N	79° 24′ W	1886–1960	600
20. Abitibi, Canada	48° 43′ N	79° 22′ W	1897–1921	500
21. New York, U.S.A.	40° 43′ N	74° 00′ W	1886–1960	850
22. Albany, U.S.A.	42° 39′ N	73° 45′ W	1886–1960	650
23. Montreal, Canada	45° 30′ N	73° 35′ W	1886–1960	800
24. Burlington, U.S.A.	44° 29′ N	73° 12′ W	1886–1960	500
25. New Haven, U.S.A.	41° 28′ N	72° 56′ W	1886–1960	850
26. Chicoutimi, Canada	48° 25′ N	71° 05′ W	1886–1960	400
27. Boston, U.S.A.	42° 21′ N	71° 04′ W	1886–1940	800
28. Father Point, Canada	48° 31′ N	68° 10′ W	1886–1950	600
29. Eastport, U.S.A.	44° 54′ N	66° 54′ W	1886–1950	600

Table 5 (continued)

Name	Latitude	Longitude	Intervalle	R_0
30. Cape Megdalene, Canada	49° 16′ N	65° 20′ W	1886–1940	500
31. South West Point, Anticosti, Canada	49° 23′ N	63° 43′ W	1886–1952	500
32. Charlottetown, Canada	46° 14′ N	63° 10′ W	1889–1960	700
33. Sable Island, Canada	43° 57′ N	60° 06′ W	1891–1960	1000
34. St. Johns, Newfoundland	47° 34′ N	52° 42′ W	1886–1940	1000
35. Madrid	40° 24′ N	3° 41′ W	1882–1960	250
36. Nantes, France	47° 15′ N	1° 34′ W	1882–1960	500
37. Paris	48° 48′ N	2° 20′ E	1882–1960	400
38. Palma (Mallorca)	39° 33′ N	2° 42′ E	1882–1960	350
39. Lyon, France	45° 41′ N	4° 47′ E	1882–1960	500
40. Trier, Germany	49° 45′ N	6° 38′ E	1882–1960	400
41. Zürich, Switzerland	47° 23′ N	8° 33′ E	1882–1960	700
42. Milano, Italy	45° 28′ N	9° 11′ E	1882–1960	700
43. Sonnblick, Austria	47° 03′ N	12° 57′ E	1892–1960	1400
44. Obir, Austria	46° 30′ N	14° 29′ E	1882–1942	1300
45. Wien, Austria	48° 15′ N	16° 22′ E	1882–1960	400
46. Krynica, Poland	49° 24′ N	20° 57′ E	1882–1960	600
47. Lwów, U.S.S.R.	49° 50′ N	24° 01′ E	1882–1960	400
48. Odessa, U.S.S.R.	46° 29′ N	30° 44′ E	1882–1960	400
49. Novorossisk, U.S.S.R.	44° 44′ N	37° 49′ E	1882–1950	400
50. Tiflis, U.S.S.R.	41° 42′ N	44° 48′ E	1882–1960	300
51. Astrachan, U.S.S.R.	46° 21′ N	48° 02′ E	1882–1960	100
52. Fort Uritsky, U.S.S.R.	44° 31′ N	50° 07′ E	1913–1950	0
53. Krasnovodsk, U.S.S.R.	40° 0′ N	52° 59′ E	1882–1960	0
54. Chimbay, U.S.S.R.	42° 57′ N	59° 49′ E	1951–1960	0
55. Turgai, U.S.S.R.	49° 38′ N	63° 27′ E	1923–1960	0
56. Tashkent, U.S.S.R.	41° 21′ N	69° 18′ E	1882–1960	200
57. Balkash, U.S.S.R.	46° 54′ N	75° 00′ E	1951–1960	50
58. Naryn Narynskoye	41° 26′ N	76° 13′ E	1882–1960	200
59. Alma Ata, U.S.S.R.	43° 16′ N	76° 53′ E	1882–1960	300
60. Mukden (Shenyang), China	41° 48′ N	123° 23′ E	1906–1940	400
61. Tunhkang, China	40° 08′ N	94° 47′ E	1938–1950	0
62. Changchun, China	43° 52′ N	125° 20′ E	1909–1944	500
63. Ghangkangin, Coree	41° 47′ N	126° 53′ E	1915–1944	500
64. Joshin, Coree	40° 40′ N	129° 11′ E	1906–1944	500
65. Unggi, Coree	42° 19′ N	130° 24′ E	1915–1945	500
66. Vladivostok, U.S.S.R.	43° 07′ N	131° 57′ E	1921–1960	500
67. Nikolsk Ussuriysky	43° 47′ N	131° 57′ E	1921–1960	400
68. Khabarorsk, U.S.S.R.	48° 31′ N	135° 10′ E	1951–1960	450
69. Nakodate, Japan	41° 49′ N	140° 45′ E	1951–1960	900
70. Aomori, Japan	40° 49′ N	140° 47′ E	1951–1960	1200
71. Sapporo, Japan	43° 03′ N	141° 20′ E	1912–1960	900
72. Wakkanai, Japan	45° 25′ N	141° 41′ E	1951–1960	900
73. Haboro, Japan	44° 22′ N	141° 42′ E	1951–1960	1200
74. Asahikawa, Japan	43° 46′ N	142° 22′ E	1951–1960	1000
75. Urakawa, Japan	42° 10′ N	142° 47′ E	1951–1960	900
76. Ochiai, Japan	47° 20′ N	142° 44′ E	1905–1944	500
77. Poronoysk, U.S.S.R.	49° 13′ N	143° 06′ E	1951–1960	500
78. Nemuro, Japan	43° 20′ N	145° 35′ E	1884–1960	700
79. Abashiri, Japan	44° 01′ N	145° 49′ E	1951–1960	1000
80. Simushir, U.S.S.R.	46° 51′ N	151° 52′ E	1951–1960	900

Table 5a

Years	$\sum_{i=1}^{N}(R_i-R_0)$	N	n	$\overline{R-R_0}$	Smoothed values	Years	$\sum_{i=1}^{N}(R_i-R_0)$	N	n	$\overline{R-R_0}$	Smoothed values
1884	3490	21		166		1923	13938	65		214	217
5	5827	21		277	237	4	12198	64		191	203
6	10860	48		226	240	5	14056	65		216	223
7	11330	49		231	247	6	17482	65		269	265
8	14461	48		301	270	7	20157	66		305	280
9	12997	53		245	254	8	15529	65		239	248
1890	12418	53		234	237	9	13693	64		210	214
1	13403	57		235	240	1930	12683	65		195	211
2	14857	58		256	247	1	15953	65		245	231
3	13861	58		239	225	2	15509	65		239	250
4	9366	57		164	201	3	17653	64		276	257
5	13647	57		239	235	4	15094	64		236	236
6	16748	56		299	276	5	16247	63		258	258
7	15175	57		266	274	6	18974	62		306	294
8	14886	56		266	255	7	18593	61		305	300
9	12462	56		223	240	8	17355	61		285	283
1900	14170	57		249	247	9	16254	63		258	260
1	15289	57		268	268	1940	15812	66		240	252
2	16612	58		287	278	1	16791	62		271	257
3	14612	57		259	258	2	14558	59		247	235
4	12985	57		228	233	3	10395	59		176	200
5	12117	56		216	216	4	11189	56		200	200
6	14932	59		258	245	5	14446	56		298	257
7	14653	59		248	242	6	11423	49	7	233	257
8	12493	58		215	242	7	13511	51	5	264	259
9	17550	61		288	256	8	13188	53		249	247
1910	13920	60		232	243	9	11214	54		226	242
1	13060	60		218	234	1950	14558	55		265	257
2	16344	61		268	246	1	15401	57		270	255
3	14063	61		231	245	2	12513	58		216	219
4	15305	61		251	256	3	12093	57		212	209
5	18624	64		291	270	4	9478	49	8	194	194
6	15186	63		241	244	5	17410	57		305	258
7	13127	64		205	205	6	13035	57		229	256
8	15652	63		249	249	7	14866	57		261	261
9	15930	60		265	253	8	16717	57		293	280
1920	13579	58		234	228	9	15301	57		273	282
1	10437	58		180	211	1960	16447	56			
2	14868	60		248	222						

The correlation between $\overline{R-R_0}$ and I_a is shown in Fig. 7, where the continuous line presents the smoothed values of $\overline{R-R_0}$ for the period 1883–1960 while the dashed line gives the corresponding values of the areas index I_a.

Table 5b

Stations	1946	1947	1954
Red. Bluff.	− 28	96	
Chicago			565
Marquette	60		
Detroit	45		
Eastport	− 31	32	
Charlottetown			579
Sonyblick	−122	−251	437
Krynica	60		
Lwów	62	142	
Sable Island		−9	
Nantes			488
Trier			371
Zürich			484
Vladivostok			438
Chicoutimi			547
$\overline{R-R_0}$	233	264	194
N	49	51	49
$\sigma\pm$	102	104	111

The crosses placed above some years present the values found for $\overline{R-R_0}$, if the irregular values of $R-R_0$ given in Table 5b are also taken into account.

As it is made clear from Fig. 7 the values of $\overline{R-R_0}$ and I_a present alternative correlations, positive and negative, during periods of time which are more or less short.

We thus get a negative correlation during the period 1885–1905.

During the period 1905–1918 the values of $\overline{R-R_0}$ present slight fluctuations round a mean value and offer insignificant correlation with the areas index I_a. During the interval 1919 to 1928 the values of $\overline{R-R_0}$ present positive correlation with the index I_a while for the next 6 years i.e. for the period 1929–1934 the correlation between these two quantities becomes negative. Finally for the last period 1935–1960 the correlation between $\overline{R-R_0}$ and I_a appears to be again positive. We thus see that during the interval 1884–1960 the correlation between $\overline{R-R_0}$ and I_a changed its sign 3 times.

The correlation coefficients between $\overline{R-R_0}$ and I_a and between $\overline{R-R_0}$ and the Wolf Numbers (W.N.) are correspondingly equal to

$$r_{\overline{R-R_0},\,I_a} = -0.62 \quad \text{Period 1885–1905, 929–1934}$$

$$r_{\overline{R-R_0},\,(\text{W.N.})} = -0.60 \quad \text{Period 1885–1905, 1929–1934}$$

$$r_{\overline{R-R_0},\,I_a} = +0.81 \quad \text{Period 1919–1928, 1935–1960}$$

$$r_{\overline{R-R_0},\,(\text{W.N.})} = +0.68 \quad \text{Period 1919–1928, 1935–1960.}$$

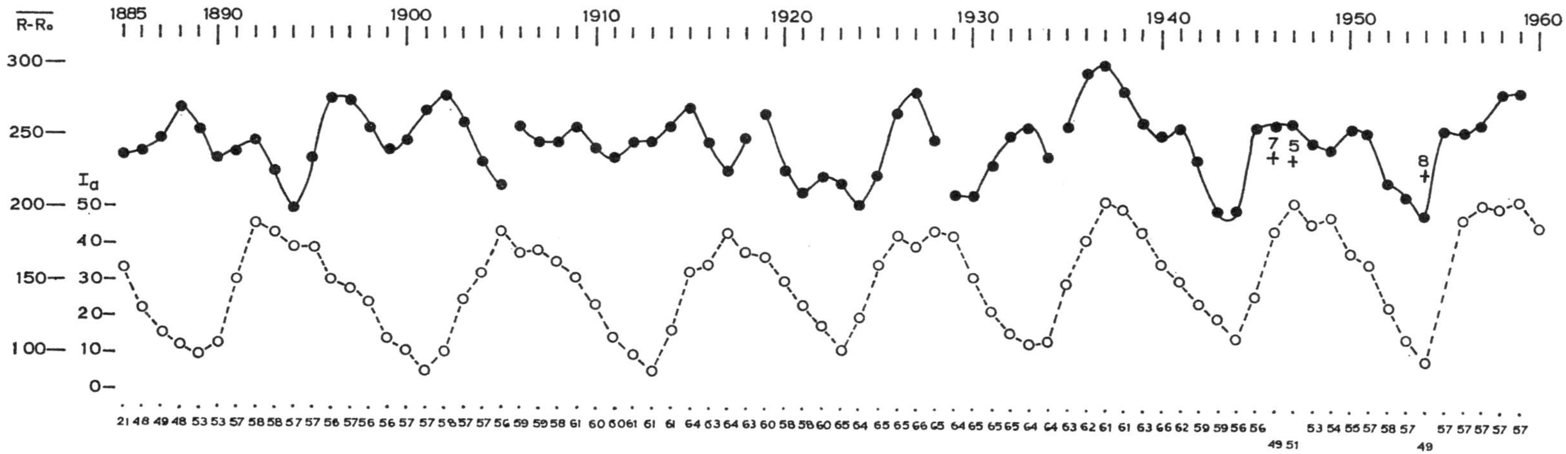

Fig. 7. Middle Latitude Zone: $40° \leqq \varphi \leqq 50°$. The continuous and the dashed line represent the mean annual variation of the quantities $\overline{R-R_0}$ and I_a respectively for the period 1883–1960. The crosses represent the values of $\overline{R-R_0}$ when some irregular values of $R-R_0$ were also used in the computation

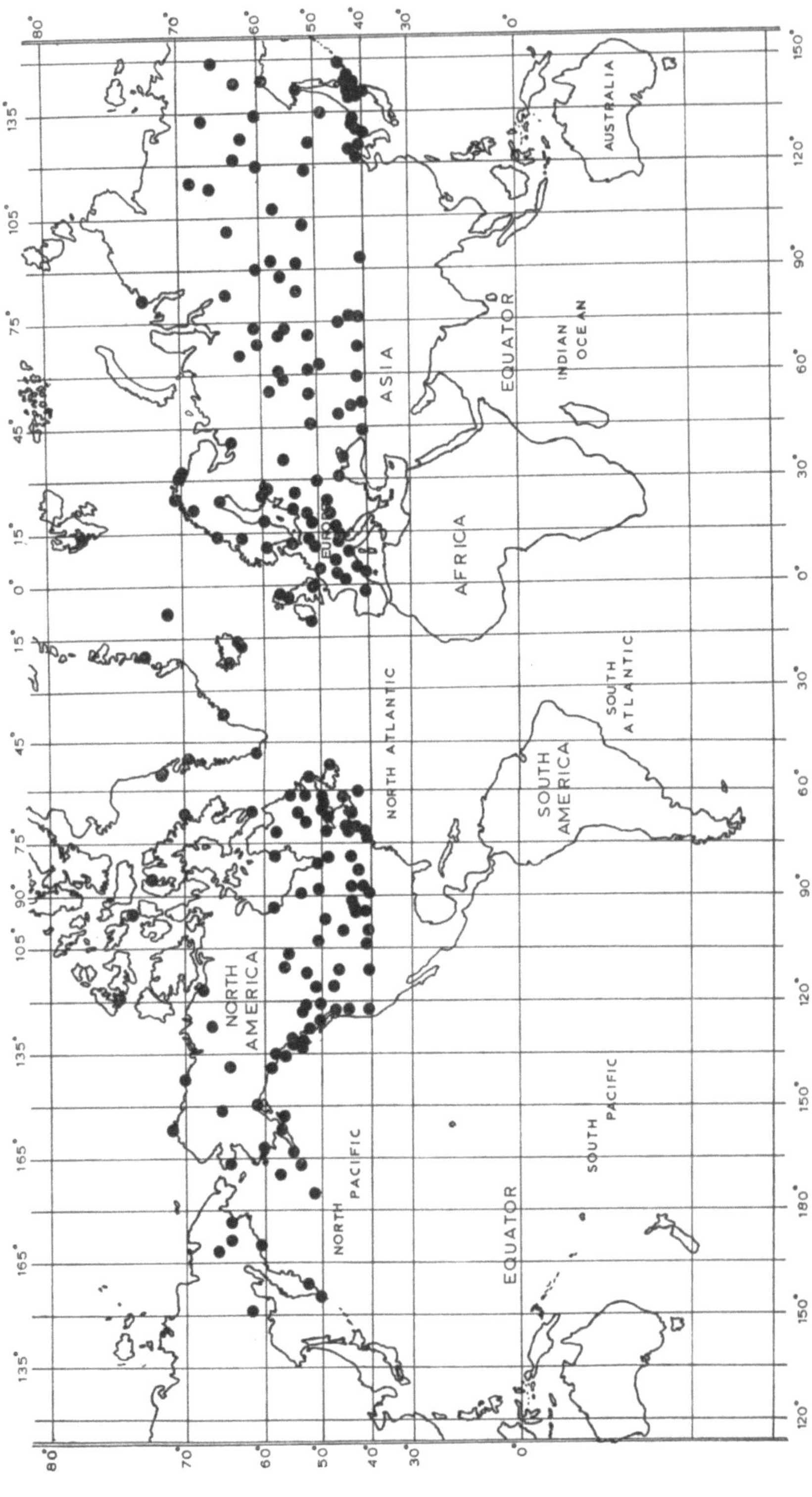

Fig. 8. Station index

If for the computation of the quantity $\overline{R-R_0}$ we also take into account the irregular values of $R-R_0$ (Table 5b) then the correlation coefficients become:

$$r_{\overline{R-R_0}, I_a} = -0.57 \quad \text{Period 1885–1905, 1929–1934}$$

$$r_{\overline{R-R_0}, (\text{W.N.})} = -0.59 \quad \text{Period 1885–1905, 1929–1934}$$

From the above results we see that the values of the correlation coefficient between $\overline{R-R_0}$ and I_a change from high to medium if for the computation of $\overline{R-R_0}$ the irregular values of $R-R_0$ are taken into account. Furthermore we see that the values of the correlations coefficient are higher if we compare $\overline{R-R_0}$ with the areas index I_a rather than with the wolf numbers (W.N.).

Conclusions

From the above discussion the following conclusions may be drawn:

1) For a given area the dispersion diagrams between the annual precipitation R and the areas index I_a present, in general, a cross-wise form.

2) The "mean zonal value of annual precipitation" $\overline{R-R_0}$, in the Polar Zone ($70° \leqq \varphi \leqq 80°$) presents a positive correlation with the areas index I_a as well as with the relative sunspot numbers.

3) In the Northern Zone the correlation between $\overline{R-R_0}$ and I_a is negative.

4) In the zone ($50° \leqq \varphi \leqq 60°$) the correlation between $\overline{R-R_0}$ and I_a is negative during the period 1890–1913 and positive during the period 1914–1960.

5) In the Middle Latitude Zone ($40° \leqq \varphi \leqq 50°$) the correlation between $\overline{R-R_0}$ and I_a changes its sign tree times, which means that it is negative during the time intervals 1885–1905, 1929–1934, and positive during the time interval 1919–1928, 1935–1960.

6) The correlation coefficients between $\overline{R-R_0}$ and I_a increase noticeably if some irregular values of $R-R_0$ are not taken into account. The percentage of these irregular values does not exceed 3% of the total number of the observations considered in each case.

The occurrence of the irregular values of the differences $R-R_0$ as well as the change of sign of the correlation from zone to zone or from period to period within the same zone should be the subject of a more extensive investigation.

References

Abbot, C.: Smiths. Miscel. Collec. Washington D. C. **117** (11) (1952).

Depietri, C.: Osservatorio Geofisico Publicazioni, Modena Università, No. 74, p. 14. 1960.

Johnson, L. P. V.: Royal Astr. Soc. Canada, Toronto, J. **99**, 250 (1964).

Papee, H., Masironi, L. A.: Geof. Pura e. Appl., Milan, **53**, 214 (1962).

Xanthakis, J.: Solar Physics, **10**, 168 (1969).

Xanthakis, J.: C. R. Acad. Sc. Paris, **271**, 1009 (1970).

Xanthakis, J.: In: C. Macris (ed.), Physics of the solar corona, p. 179. Rheidel Publ. Co. 1970.

Discussions

Kiepenheuer, K. O.:

I think it would be quite a valuable addition if you would discuss your correlation results also by comparing them with the ionospheric data (mostly critical frequencies) obtained in the terrestrial areas, when precipitation was recorded.

Xanthakis, J.:

I made no such comparison because the ionospheric data refer only to the latest years, and cover a short period of time. The above investigation on rainfall covers from the period 1882 until now.

A Study of the Areas Index I_a

By C. POULAKOS and B. TRITAKIS
Research Center for Astronomy and Applied Mathematics, Academy of Athens, Greece

With 8 Figures

Abstract

In the first part of this paper, beyond a comparative study of the variation of the areas index I_a, the relative sunspot numbers R and the areas of sunspots A, we propose a relation of the form:

$$[I_a]=b(11-T_R)\cos\frac{\pi}{2\Omega_{1,2}}t+F_{1,2}$$

where $[I_a]$ represents the mean annual values of the areas index I_a, T_R is the time of rise corresponding to the relative sunspot numbers, $F_{1,2}$ is a periodic term and b is a constant.

In the second part a relation of the form:

$$[I_a]=C\sqrt{[R]}+P$$

is proposed, where $[I_a]$ represents the mean annual values of the areas index I_a, $[R]$ represents the corresponding values of the Wolf numbers, P is a periodic term and C is a constant.

In the third part the seasonal variation of the areas index I_a, as well as of the relative sunspot numbers R, is studied for the period 1880 to 1964.

I. Introduction

It is generally accepted that the main phenomena characterizing the variation of solar activity within each sunspot cycle and from cycle to cycle, are the sunspots and the faculae. It is, therefore, preferable for the study of the long-term variation of solar activity to use, instead of the relative sunspot numbers R, the recently introduced (XANTHAKIS, 1969) new index of solar activity which could express better one principal characteristic of an active region, i.e. the areas index I_a defined by the relation (XANTHAKIS, 1969):

$$I_a=\tfrac{1}{2}[\sqrt{A}+\sqrt{f}] \tag{1}$$

where A and f are correspondingly the total areas of sunspots and faculae corrected for foreshortening. These are given by the relevant publications of Greenwich Observatory.

As it is seen from equation 1 the areas index I_a may be more complicated when compared with the Wolf numbers R but, on the other part, I_a presents many advantages over R. This is due to the fact that the areas index I_a depends directly on the areas of sunspots and faculae i.e., on two solar phenomena which, in turn, depend to a great extent, on the solar magnetic field strength.

It is also worth mentioning that some indices of solar activity, such as the intensity of the solar-radio emission S_θ in the frequency range $1000\ \text{MHz} \leqq f \leqq 10000\ \text{MHz}$, the intensity of the coronal line I_{5303}, the flare index I_f etc., are more closely correlated with the areas index I_a than with the relative sunspot numbers R and in some cases the combination of I_a with the above mentioned indices of solar activity provides interesting relations (XANTHAKIS, 1970). Besides, as it has recently been shown by XANTHAKIS, the correlation coefficient between the mean annual variation of the precipitation $\overline{R-R_0}$ and the areas index I_a is by far superior to that between the quantity $\overline{R-R_0}$ and the relative sunspot numbers.

In view of these facts we consider that it is preferable to use the areas index I_a as an index of solar activity, instead of the relative sunspot numbers.

In what follows, apart from a comparative study of some indices of solar activity, an attempt has been made to find the analytical expression of the areas index I_a as a function of the time of rise T_R and as a function of the relative sunspot numbers R.

The seasonal variation of the areas index I_a as well as of the relative sunspot numbers R is studied for the period 1880–1964.

However, as values of the areas of sunspots A and faculae f do not exist for the sunspot cycles preceding cycle No. 12, the discussion is based upon the sunspot cycles $N=12$ to 19.

II. The Areas Index I_a and the Time of Rise T_R

Before taking up the main subject we may attempt a comparison of the variation of the areas index I_a with the variation of the areas of sunspots and of the relative sunspots numbers, shown in Fig. 1.

From Fig. 1 we see that for all the solar cycles of the time interval considered here, the curve corresponding to the mean annual values of the areas index I_a is, by far, smoother than the curves corresponding to the mean annual values of the areas of sunspots and of the relative sunspots numbers, while at the same time, it does not miss any of those details showing the fluctuation of solar activity. It is also worth-noting that this feature is more characteristic for cycles Nos. 17, 18 and 19. In any case, the three curves of Fig. 1 offer the same trends, a fact which renders obvious a strong correlation among these indices of solar activity.

As has already been shown (XANTHAKIS, 1967) the different indices of solar activity can be analytically expressed as function of only one parameter i.e. the time of rise T_R corresponding to the relative sunspot numbers. We considered then, that it would be interesting to examine whether a similar dependence exists in the case of the areas index. As a result of this investigation we found that the values $[I_{a_m}]$ i.e. the mean annual values of I_a corresponding to the years of sunspot

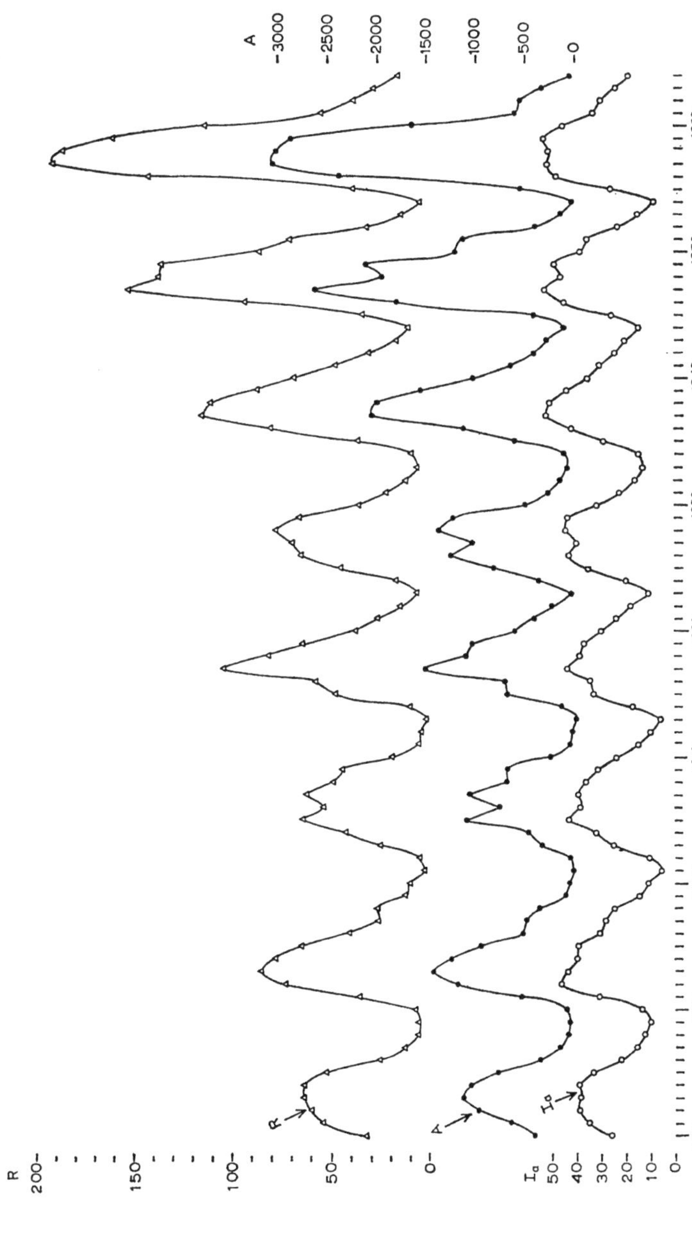

Fig. 1. A comparison of the mean annual variation of the areas index I_a with the corresponding values of the areas of sunspots A and of the relative sunspot numbers R, for the period 1880–1964(N= 12 to 19)

Table 1. The observed and computed values of $[I_{a_m}]$ corresponding to the years of sunspot maximum

N	12	13	14	15	16	17	18	19
$[I_{a_m}]^{ob}$	38.57	43.64	38.80	43.60	43.84	52.37	52.58	51.48
$[I_{a_m}]^{c}$	39.33	44.16	35.88	46.92	42.09	50.37	53.82	53.13
$o-c$	−0.76	−0.52	+2.92	−3.32	+1.75	+2.00	−1.24	−1.65
T_R	5.3	4.6	5.8	4.2	4.9	3.7	3.2	3.3

maximum, can be analytically expressed as a function of the time of rise T_R by means of the relation:

$$[I_{a_m}] = b(11 - T_R) \tag{2}$$

where $b = 6.9$.

As is shown in Table 1, Eq. (2) represents the observational values of the quantity I_{a_m} with a very satisfactory approximation.

The accuracy was calculated with the help of the formula $\left(1 - \frac{\sigma}{[I_{a_m}]}\right) 100\%$ and has been found equal to 96%.

It is however a striking fact that relation 2 is a linear function of the time of rise T_R.

The variation of the mean annual values of $[I_a]$ within each sunspot cycle can also be expressed analytically function of the same parameter T_R with the help of the relation:

$$[I_a] = [I_{a_m}] \cos \frac{\pi}{2\Omega_{1,2}} t + F_{1,2} \tag{3}$$

where $\Omega_1 = T_R$ for the ascending branch
$\Omega_2 = 11 - T_R$ for the descending branch
$t = 1, 2, 3, \ldots$ for the first, second, third etc. year preceding or following the year of sunspot maximum.

The terms F_1, and F_2 are periodic terms and correspond respectively to the ascending and the descending branch of each sunspot cycle, given by the relation:

$$\left.\begin{aligned} F_1 &= 2b \sin N \frac{\pi}{6} \sin \frac{\pi}{6} t - 5 \sin(N+2) \frac{\pi}{8} \sin \frac{\pi}{3} t \\ F_2 &= -5 \cos(N+1) \frac{\pi}{6} \sin \frac{\pi}{5} t \end{aligned}\right\} \text{even cycles} \tag{4}$$

$$\left.\begin{aligned} F_1 &= -5 \sin \frac{\pi}{3} (t-1) \\ F_2 &= b \left[\sin \frac{\pi}{10} (t-1) - 5 \sin \frac{\pi}{5} (t-1)\right] \end{aligned}\right\} \text{odd cycles} \tag{5}$$

where $b = 6.9$

Relation 3, which is, in fact, quite similar to that proposed recently by XANTHAKIS (1969), represents the observational data with an accuracy equal to 89%.

III. The Areas Index I_a and the Wolf Numbers R

As has already been shown by XANTHAKIS (1969) the values of $[R]$ i.e. the mean annual values of the relative sunspot numbers R can be expressed analytically as a function of the time of rise T_R with the help of the relation:

$$[R]=[R_m]\cos^2\frac{\pi}{2\Omega_{1,2}}t+G_{1,2} \tag{6}$$

where $[R_m]$ is a polynomial

$$[R_m]=C_R+2b_R(T_R-T_0)^2+b_R Y \tag{7}$$

of the second degree with respect to the parameter T_R. Taking now into consideration that relation 3, which gives the mean annual values of the areas index $[I_a]$ as a function of the time of rise, is a linear function of T_R, it results from Eqs. (3), (6) and (7) that the areas index $[I_a]$ is a linear function of $\sqrt{[R]}$.

To this end we calculated the correlation between the observed mean annual values of $[I_a]$ and the corresponding values of $\sqrt{[R]}$ for the sunspot cycles for which observational data were available i.e. for cycles $N=12$ to 19.

The numerical results are shown in Fig. 2.

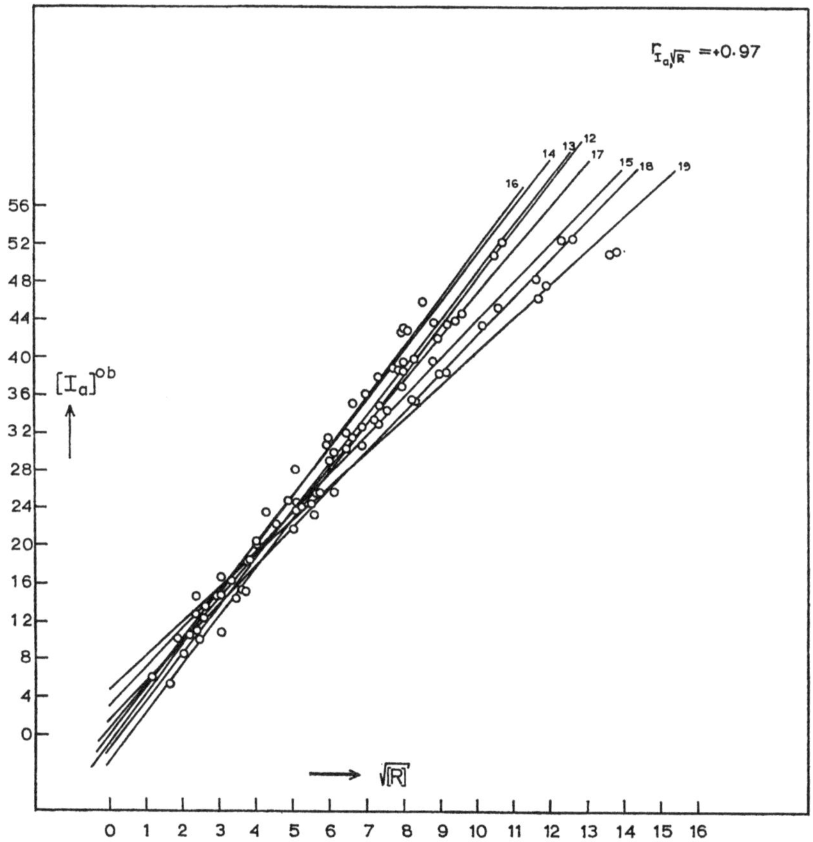

Fig. 2. Correlation between the mean annual values of $[I_a]^{ob}$ and the corresponding values of the quantity $\sqrt{[R]}$ for each of the sunspot cycles $N=12$ to 19. In the upper part the correlation coefficient $r_{I_a,\sqrt{R}}$ for all the years considered is noted

From the calculations we found that the analytical expression of the areas index $[I_a]$ as a function of $\sqrt{[R]}$ is of the form:

$$[I_a]=C\sqrt{[R]} \tag{8}$$

where $C=4.65$.

If we study the differences $[I_a]^{ob}-C\cdot\sqrt{[R]}$ we see that while these differences are small for the years corresponding to the ascending and the descending branch of each cycle they become quite high during the years of sunspot maximum. The variation, on the other hand, of these differences, as shown in Fig. 3, appears to be periodic in character with a period of about 5 years, a fact which led us to find a periodic term of the form:

$$P=5\left[\sin N\frac{\pi}{2}+\frac{1}{2}\sin\left(N\frac{\pi}{5}-\frac{\pi}{4}\right)\right] \tag{9}$$

which, when applied to the values of $[I_{a_m}]^c$, significantly reduces the discrepancies between the observed and computed values of $[I_{a_m}]$.

Consequently from Eqs. (8) and (9), it follows that the mean annual values of the areas index $[I_a]$ can be analytically expressed as a function of $\sqrt{[R]}$ by means of the relation:

$$[I_a]=C\sqrt{[R]}+P \tag{10}$$

where P is active only for the years of sunspot maximum.

The above results are better shown in Fig. 3, where the continuous line illustrates the values of $[I_{a_m}]^{ob}-C\sqrt{[R]}$ and the dashed line the corresponding differences properly corrected by means of the periodic term P.

Table 2 gives the observed mean annual values $[I_a]^{ob}$ of $[I_a]$ as well as the computed values $[I_a]^c$ of this quantity through Eq. (10) and the corresponding differences $o-c=[I_a]^{ob}-[I_a]^c$ for the time interval 1880–1964.

From Table 2 it follows that Eq. (10) represents the observational data with an accuracy equal to 89.6%. This accuracy becomes fairly higher if we exclude the values of $[I_a]$ corresponding to the years surrounding the year of sunspot maximum of the "anomalous" cycle No. 19 for which Eq. (10) assumes high values of $[I_a]$ for these years. This slight discrepancy between the observed and computed values of $[I_a]$ is shown in Fig. 4. No doubt both curves in this figure show offer considerable similarity.

Finally we considered that it would be of some interest to examine whether relation 10 could also be extended to the case of the mean quarterly and the mean monthly values of the areas index I_a.

As it results in fact from Fig. 5 and the accuracy for the mean monthly values of I_a, which has been found equal to 88%, Eq. (10) is also valid for these values of I_a.

IV. Seasonal Variation of the Areas Index I_a and the Wolf Numbers R

It has recently been shown by FRITZOVÁ-ŠVESTKOVÁ and ŠVESTKA (1966), that in northern winter months there is a significant minimum in the seasonal distribution of the occurrence frequency for PCA while, as we can see in Fig. 6 a maxi-

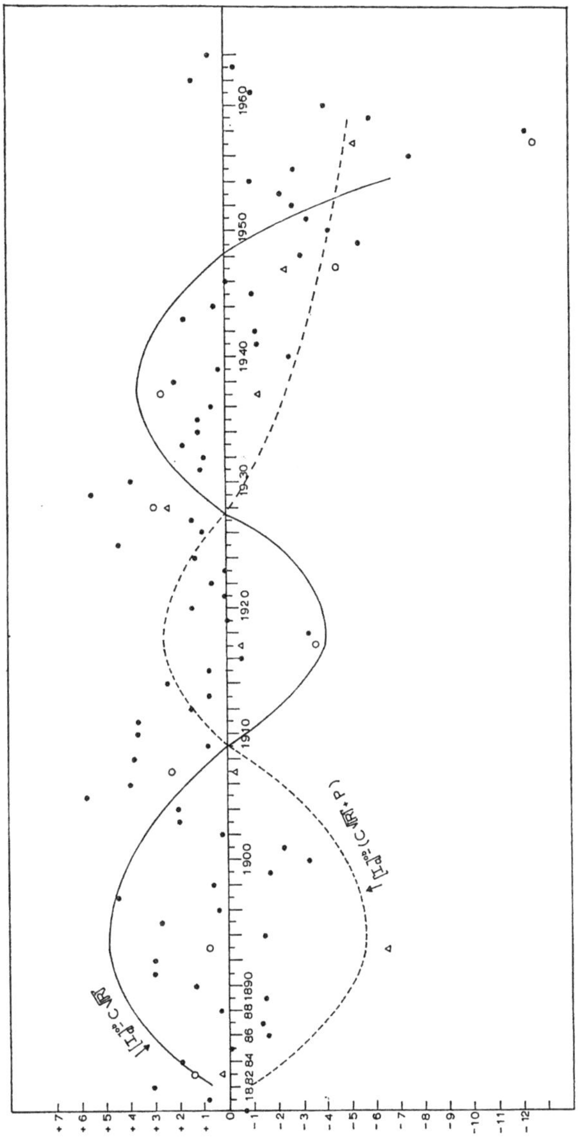

Fig. 3. The continuous line presents the differences $[I_{a_m}]^{ob} - C\sqrt{[R]}$ and the dashed line represents the differences $[I_{a_m}]^{ob} - (C\sqrt{[R]} + P)$. The black circles represent the differences $[I_{a_m}]^{ob} - C\sqrt{[R]}$

mum is noticeable in the summer months i.e. in the third quarter of the years 1956–1963.

Furthermore, XANTHAKIS has recently shown (1969), that during the sunspot cycle No. 19 the seasonal variation of the number of proton flares N_{PF}, presents its maximum in the third quarter of the considered years (1955–1964) as it is seen in Fig. 7.

After having gone over the cited papers of ŠVESTKA and XANTHAKIS we considered that it was worthwhile to examine whether the areas index I_a and the relative sunspot numbers R present a similar seasonal variation.

Table 2. Mean annual values of the areas index $[I_a]^{ob}$ as well as the values of $[I_a]^{c}$ computed with the help of relation 10 and the corresponding differences $o-c$

Year	$[I_a]^{ob}$	$[I_a]^{c}$	$o-c$	Year	$[I_a]^{ob}$	$[I_a]^{c}$	$o-c$
1880	25.69	26.41	−0.72	3	11.16	11.21	−0.05
1	35.10	34.27	+0.83	4	20.30	19.02	+1.28
2	39.01	35.94	+3.07	5	35.32	30.97	+4.35
3	38.57	38.25	+0.22	6	43.04	37.15	+5.89
4	38.98	37.06	+1.92	7	39.78	38.64	+1.14
5	33.48	33.57	−0.09	8	43.84	41.40	+2.44
6	21.78	23.44	−1.66	9	42.96	37.48	+5.48
7	15.41	16.83	−1.42	1930	31.54	27.76	+3.78
8	12.44	12.14	+0.30	1	22.44	21.39	+1.05
9	10.14	11.67	−1.53	2	16.38	15.48	+0.90
1890	13.70	12.37	+1.33	3	12.86	11.11	+1.75
1	30.74	27.76	+2.98	4	14.86	13.72	+1.14
2	46.00	39.71	+6.29	5	29.08	27.95	+1.13
3	43.64	50.10	−6.46	6	42.12	41.52	+0.60
4	39.56	41.06	−1.50	7	52.37	53.63	−1.26
5	39.47	37.20	+2.27	8	50.77	48.69	+2.08
6	30.42	30.09	+0.33	9	44.10	43.80	+0.30
7	28.28	23.81	+4.47	1940	35.64	38.27	−2.63
8	24.60	24.04	+0.56	1	30.77	32.04	−1.27
9	14.45	16.18	−1.73	2	24.50	25.71	−1.21
1900	11.04	14.32	−3.28	3	20.50	18.79	+1.71
1	5.43	7.63	−2.20	4	14.88	14.42	+0.46
2	10.60	10.42	+0.18	5	25.68	26.78	−1.10
3	24.92	22.97	+1.95	6	44.70	44.73	−0.03
4	32.10	30.13	+1.97	7	52.58	55.01	−2.43
5	42.81	37.06	+5.75	8	46.37	54.27	−7.90
6	38.03	34.08	+3.95	9	48.55	53.99	−5.44
7	38.80	39.07	−0.27	1950	38.42	42.59	−4.17
8	36.10	32.36	+3.74	1	35.42	38.73	−3.31
9	31.58	30.83	+0.75	2	23.38	26.09	−2.71
1910	23.70	20.04	+3.66	3	15.14	17.34	−2.20
1	14.71	11.11	+3.60	4	8.77	9.77	−1.00
2	10.28	8.84	+1.44	5	25.85	28.64	−2.79
3	6.20	5.49	+0.71	6	47.89	55.34	−7.45
4	16.82	14.42	+2.40	7	51.48	56.65	−5.17
5	32.70	31.99	+0.71	8	51.04	63.19	−12.15
6	34.58	35.15	−0.57	9	52.80	58.64	−5.84
7	43.60	44.15	−0.55	1960	45.32	49.29	−3.97
8	38.41	41.76	−3.35	1	33.10	34.13	−1.03
9	37.00	37.06	−0.06	2	29.90	28.46	+1.44
1920	29.88	28.50	+1.38	3	24.20	24.55	−0.35
1	23.84	23.76	+0.08	4	18.60	17.90	+0.70
2	18.12	17.53	+0.59				

To this end and in order to study this phenomenon in detail we considered the variation of the index I_a as well as of the Wolf numbers R, separately for each sunspot cycle, of cycles $N=12$ to 19 by means of the mean quarterly values of these two quantities.

The seasonal variation of the areas index I_a as well as of the relative sunspot numbers R, for the even sunspot cycles Nos. 12, 14, 16 and 18 is shown in Fig. 8a,

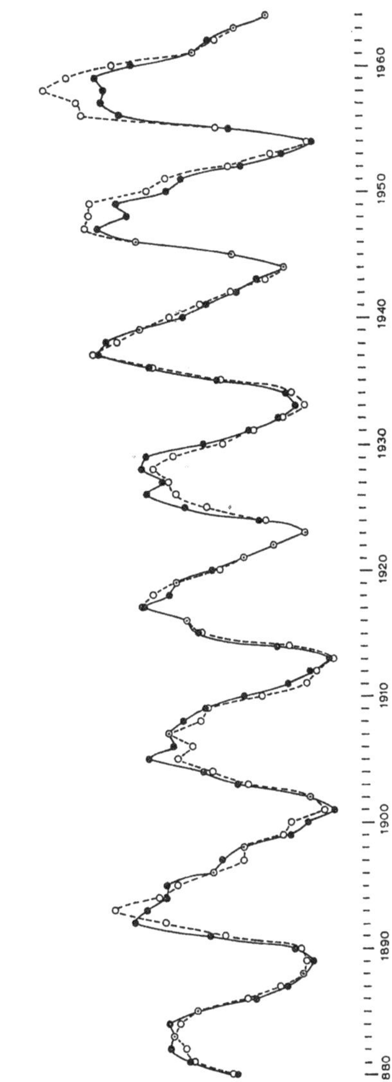

Fig. 4. The continuous line illustrates the observed mean annual values of the areas index $[I_a]$ and the dashed line represents the values of $[I_a]$ computed with the help of Eq. (10)

where the open circles represent the mean quarterly values of the areas index I_a while the black circles represent the corresponding values of the relative sunspot numbers.

From Fig. 8a we see that there is no systematic tendency of increasing or decreasing values of I_a and R, a fact which leads to the conclusion that during the even sunspot cycles considered here, there is not any characteristic seasonal variation of these two quantities.

On the contrary, as it is shown in Fig. 8b, in which the symbols are similar to those of Fig. 8a, the areas index I_a and the relative sunspot numbers R present

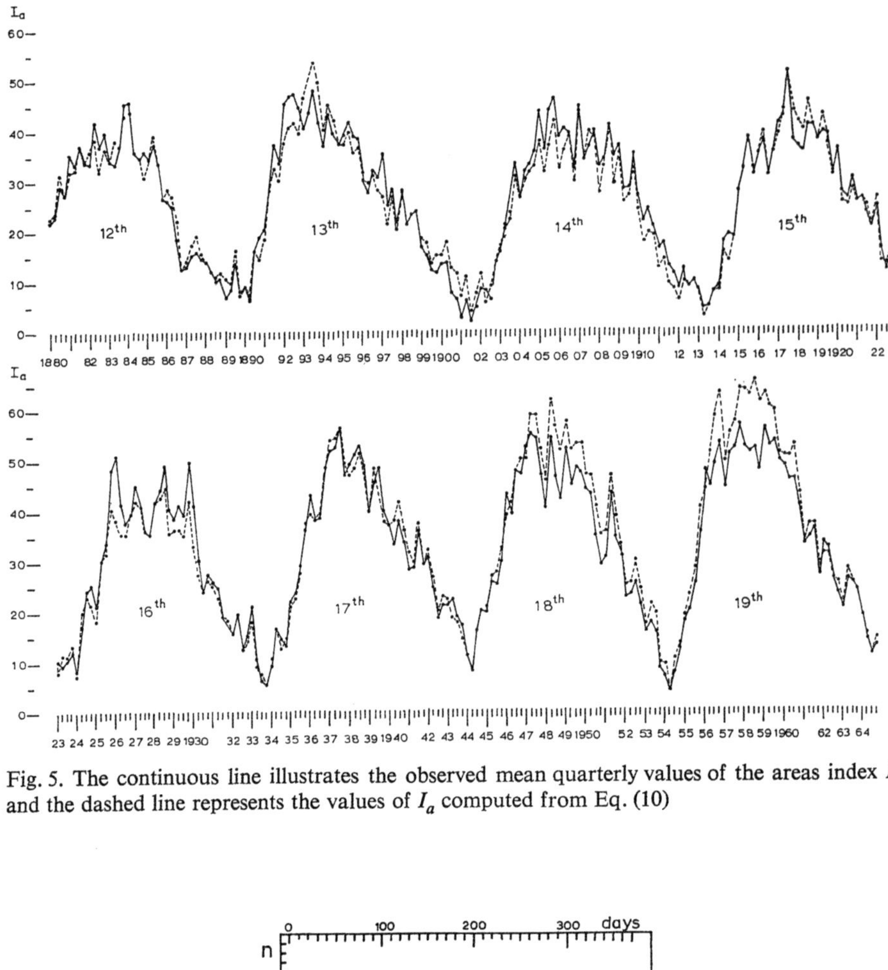

Fig. 5. The continuous line illustrates the observed mean quarterly values of the areas index I_a and the dashed line represents the values of I_a computed from Eq. (10)

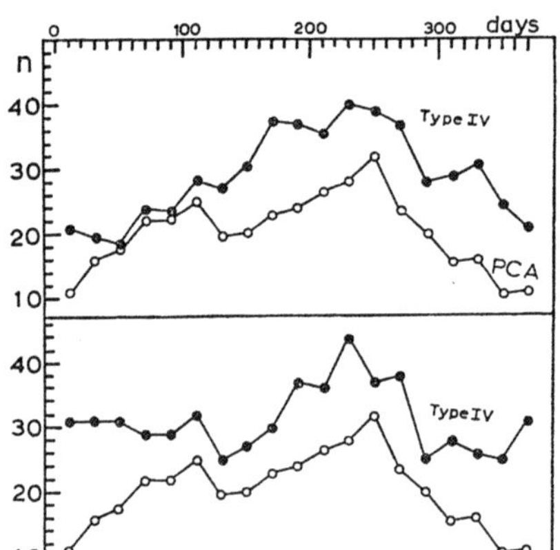

Fig. 6. Observed seasonal variation of flares of type IV radio bursts and of PCA (upper part). In the lower part the type IV burst occurrence has been corrected for the incomplete radio patrol in northern winter months. Individual points are sums within 60 days

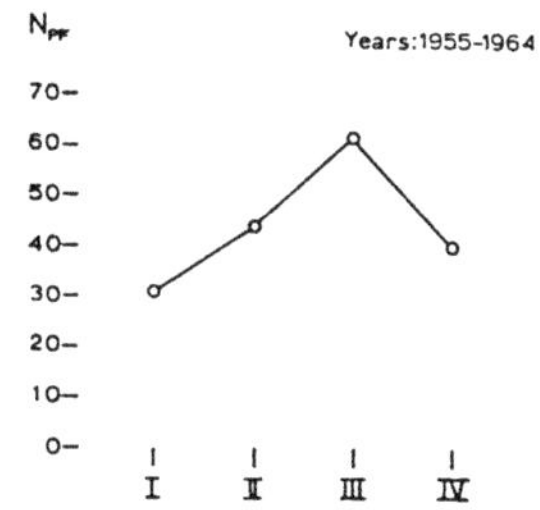

Fig. 7. Seasonal variation of the number of proton flares N_{PF} for the 19th sunspot cycle

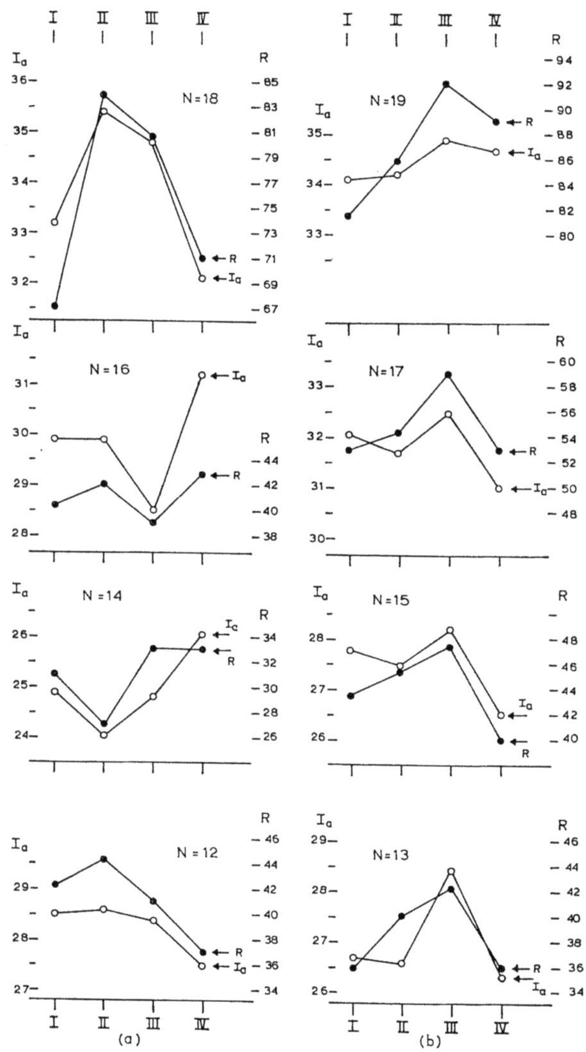

Fig. 8. Seasonal variation of the areas index I_a and of the relative sunspot numbers for the sunspot cycles $N=12$ to 19

during the odd sunspot cycles Nos. 13, 15, 17 and 19, a considerable seasonal variation which appears to have its maximum during the third quarter of the considered years in each sunspot cycle.

The above results coincide quite well with those found by Švestka and Xanthakis as concerns the PCA and the number of proton flares respectively.

References

Fritzová-Švestková, L., Švestka, Z.: Bull. Astron. Inst. Czech. (BAC) **17**, 249 (1966).
Waldmeier, M.: The sunspot activity in the years 1610–1960, Zurich: Schultes, 1961.
Xanthakis, J.: *Solar physics*, p. 157–227. London: Interscience Publishers J. Wiley & Sons.
Xanthakis, J.: Solar Physics **10**, 168 (1969).
Xanthakis, J.: In: C. Macris (ed.), Physics of the solar corona, vol. 27, p. 179–191. Reidel Publishing Co. 1970.

Discussion

Zerefos, C. S.:
The areas index I_a shows a double peak structure in each solar maximum, although the second peak is smaller.

Poulakos, C.:
This is correct for I_a and furthermore one of the advantages of this new index of Solar Activity which, in spite of the one peak showed by the Wolf numbers, present, however, a double peak during the maximum of the 19th sunspot cycle.

The Solar Activity Cycle and the Breathing of the Heliosphere

By H. J. FAHR
Institut für Astrophysik und Extraterrestrische Forschung der Universität Bonn
Bonn, Germany

With 6 Figures

Abstract

Interstellar helium and hydrogen enter the solar system from a specific direction of the interstellar space and suffer loss processes due to ionization and charge exchange reactions. The geometry of the cavity of ionized matter is determined by the effects of macroscopic streaming, extinction along hyperbolic orbits and gravitational focusing. Whereas the helium cavity turns out to be similar to a sphere which is strongly cusped in the direction opposite to the approach of interstellar matter, the hydrogen cavity resembles a sphere largely elongated in that direction. This difference in the geometries of the cavities is due to the enormously different radiation pressures acting upon neutral helium and hydrogen. While solar radiation pressure is negligible for helium, it nearly compensates for the gravitational forces in the case of hydrogen.

Since both the solar EUV-fluxes below 912 and 504 Å and the solar resonance line fluxes at 126 and 584 Å vary appreciably during the 11-year solar activity cycle, both geometries and dimensions of the helium and hydrogen cavities show pronounced solar cycle-correlated variations. Connected with these variations are variations of the solar wind dynamics, especially of the solar wind kinetic energy flux in the outer heliospheric regions. This flux variation, however, makes the heliopause, the boundary where the radially expanding solar wind terminates, vary with the solar cycle. It is shown that solar cycle heliopause variations, with respect to corresponding cavity variations, are comparatively small.

I. Introduction

It has been shown that neutral interstellar matter is able to approach the sun up to solar distances r_c of the order of a few AU, before ionization due to solar EUV-photons or charge-transfering solar wind protons occurs (BLUM and FAHR 1970, HOLZER and AXFORD 1971). This deep penetration into the solar system can take place, since the interstellar medium that surrounds the solar system is in a macroscopic motion v_0 of the order of 20 km/sec relative to the sun. As collisional forces acting upon neutral atoms in interplanetary space can be neglected beyond a specific solar distance $d_c \ll r_c$, the motion of interstellar neutrals is described by hyperbolical orbits through the solar gravitational field. Extinction losses E_1 due to photoionization and E_2 due to charge transfer processes occur, while the neutrals move along their Keplerian orbits. On the other hand, the gravitational focusing E_0 counteracts these extinction losses and even overcompensates them in the downwind regions of the interplanetary space opposite to the direction $(\boldsymbol{v}_0/v_0)$ of approach of interstellar matter.

The result of the macroscopic motion $\boldsymbol{v}_0$ of interstellar matter is an anisotropic density distribution in interplanetary space given by:

$$n(r, \theta) = n_0 \cdot E_0(r, \theta) \cdot E_1(r, \theta) \cdot E_2(r, \theta) \tag{1}$$

where n_0 is the density of the neutral interstellar constituent, i.e. hydrogen or helium, outside the solar system, r is the solar distance, and θ the angle between the vector $\boldsymbol{r}$ and the velocity $\boldsymbol{v}_0$. The functions $E_{0,1,2}$ have been given by BLUM and FAHR (1970). For interstellar temperatures $T_i \geqq 1000$ K or for the downwind regions where gravitational focusing is important, the thermal velocities $\boldsymbol{w}$ superimposed on $\boldsymbol{v}_0$ must be taken into account in the calculation of $n(r, \theta)$. The general theory and calculations for this case $w \gtrless v_0$ have been given by FAHR (1971a). A low temperature approximation $(w \ll v_0)$ has also been presented by FELDMAN *et al.* (1971) at the Asilomar conference.

II. Solar Cycle Correlated Radiations

The interplanetary density of interstellar hydrogen and helium depends both on the intensity of the ionizing solar radiations and on the net solar gravitational force. The ionization of neutral hydrogen is mainly due to the flux of solar EUV-photons below 912 Å and the solar wind proton flux, whereas the impact ionization by solar electrons can be neglected. The ionization of helium atoms is mainly due to the flux of solar EUV-photons below 504 Å, whereas charge exchange with solar wind protons and α-particles gives only a five-percent correction (see for instance AXFORD, 1971).

The solar EUV-flux has turned out to be solar-cycle-correlated with a sufficient confidence. Former results of HINTEREGGER *et al.* (1963) have led HARRIS and PRIESTER (1962) to formulate a general correlation between the smoothed values of the 10.7-cm-radiation and the EUV-radiation of the sun. Since the correlation of the 10.7-cm flux with the Zurich sunspot number R_z is well-established (SLUTZ *et al.*, 1971), this fact seems to support a sufficiently pronounced solar cycle variation of the EUV-flux. This is also confirmed by more recent observational results of HALL *et al.* (1969) who have found a general linear trend for higher EUV-fluxes to occur in correlation with higher 10.7-cm fluxes, though a strict phase correlation is not founded. However, in order to give a survey on effects triggered by flux variations during the solar activity cycle, we can adopt the linear-trended correlation given by HALL *et al.* (1969) as an indication for the mean EUV-flux variations both for photons below 912 Å and below 504 Å.

A correlation of the solar wind parameters to the solar activity cycle is much less clear. This seems mainly due to the fact that the solar wind primarily is a quiet-disk phenomenon, whereas the activity centers rising in number during the 11-year solar cycle contribute only a superposition of high-velocity plasma which is of minor importance for the average solar wind parameters. Looking over all solar wind data available so far, we find no clear evidence for a solar cycle correlation of the solar wind parameters. In the data of the satellites Pioneer VI and VII that covered two different time periods with different solar activities an indication for the occurrence of higher bulk velocities at higher solar activities R_z seems to be given. Pioneer VI, which was active from Dec. 18, 1965 to Feb. 15, 1966 when the mean activity was $\bar{R}_z=20$, gave an average bulk velocity $\bar{v}_{sw}=420$ km/sec, whereas Pioneer VII, active from Aug. 15 to Oct. 15, 1966, when the mean activity was $\bar{R}_z=50$, showed this velocity to be $\bar{v}_{sw}=450$ km/sec (MIHALOV and WOLFE, 1971). On the other hand both Pioneers have shown up a slight anticorrelation of the solar wind proton flux f_p and the bulk velocity v_{sw} (HIRSHBERG *et al.*, 1972) which is, however, negligibly pronounced for velocities above $v_{sw}=400$ km/sec. Therefore we can state that the average solar wind proton flux $\bar{f}_p$ which determines the extinction losses E_2 for neutral hydrogen due to charge exchange processes has no evident solar cycle variation.

ROBBINS *et al.* (1970) have found that the mean helium-to-hydrogen abundance ratios (He^{++}/H^{+}) in the solar wind, as obtained by the satellites Vela 3A and Vela 3B, has a slight correlation with the sunspot number R_z during the period 1965–1967. The simultaneous occurrence of high helium abundance ratios and high solar wind velocities v_{sw} could in all cases be related to the existence of a large center of activity and the occurrence of several large flares. This fact led HUNDHAUSEN (1970) to the conclusion that the occurrence of solar flares, which usually produce interplanetary shock waves, is a necessary condition for anomalously high helium abundances in the solar wind. It is thus clear that since there exists a positive R_z-correlation both of $\bar{v}_{sw}$ and the density $\bar{n}_\alpha$ of α-particles in the solar wind, the flux of solar wind α-particles $\bar{f}_\alpha=\overline{n_\alpha v_{sw}}$ is also correlated to R_z. The flux f_α contributes to the charge exchange losses of neutral helium. Since, however, photoionization of helium is the largely dominant loss process, the solar cycle variation of f_α has a negligibly small triggering effect on the density $n_{He}(r, \theta)$ of interstellar helium in interplanetary space.

So far we can thus conclude that the densities $n_H(r, \theta)$ and $n_{He}(r, \theta)$ of helium and hydrogen are triggered only by the functions $E_{1,H}$ and $E_{1,He}$ which vary with the solar-cycle-correlated EUV-fluxes.

The function $E_{0\,H}$ which describes the focusing effect of the solar gravitational field on hydrogen atoms also varies during the solar cycle, since the net gravitational force K is activity-correlated by the solar Lyman-Alpha emission. This is due to the fact that the solar Lyman-Alpha radiation is scattered by neutral hydrogen and thereby exerts a pressure K_s on it which is in antisolar direction. This pressure K_s can be described as a repulsive force decreasing with the square of the solar distance r and partly compensating the gravitational attraction K_g. Whereas the solar radiation pressure at 584 Å acting upon neutral helium is completely negligible; it compensates for 1/3 to 2/3 of the gravitational force on hydrogen, depending on the intensity of the solar Lyman-Alpha radiation. (See for example FAHR, 1971a). This is calculated by the use of two intensity-calibrated high-resolution-profiles of the solar Lyman-Alpha line measured by TOUSEY (1963) for Aug. 22, 1962 (near solar minimum conditions) and July 21, 1959 (near solar maximum conditions). The effect of a solar Lyman-Alpha radiation pressure vary-

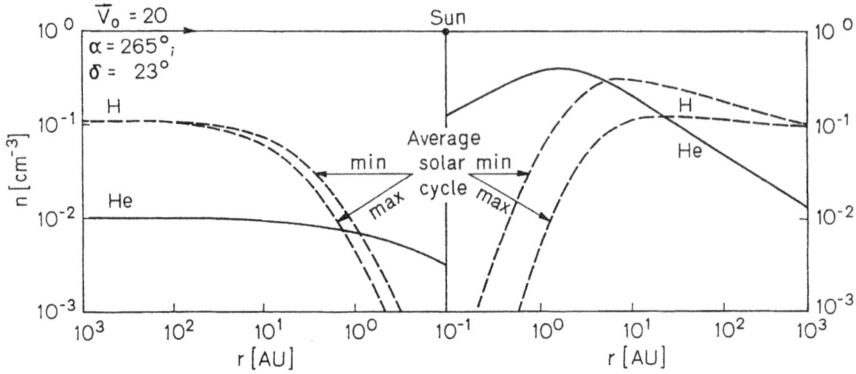

Fig. 1. The densities of interstellar hydrogen and helium are shown versus the distance r from the sun in upwind ($\theta=0°$) and downwind ($\theta=180°$) directions. Whereas the helium profiles are practically not affected by the solar 584-Å-radiation pressure, the hydrogen profiles change appreciably with the change of the solar Lyman-Alpha radiation pressure. Calculations are based on measurements reviewed by TOUSEY (1963). v_0 has been assumed to be 20 km/sec, the temperature to be $T=250$ K

ing in this range on the density distribution of hydrogen in the interplanetary space is shown in Fig. 1 for both the downwind and the upwind directions.

Recently BLAMONT and VIDAL-MADJAR (1971) have shown that the central solar line intensity at 1216 Å, which is relevant for the actual radiation pressure, since absorption occurs near the center of the line, has a pronounced trend to vary with the sunspot number R_z. This trend is shown up in a linear least-square fit of the data which gives a maximum-to-minimum variation I_{max}/I_{min} of about 2.4 between $0 \leq R_z \leq 200$ which is the range representative for a typical solar cycle activity variation. If we consider the high resolution profile for Aug. 22, 1962 of TOUSEY (1963) to be representative for a typical solar minimum profile of the Lyman-Alpha line, we are led to a radiative compensation of about 33 percent of the solar gravity on neutral hydrogen during solar minimum conditions. We will take this range as typical for the variation of the solar Lyman-Alpha radiation pressure over a solar cycle, though cases of even total compensation of gravity may have been registrated during 1969 (BLAMONT and VIDAL-MADJAR, 1971; THOMAS, 1972).

Table 1. Standard solar cycle

	Solar minimum conditions	Solar maximum conditions
Photoionization rate ν_H	$0.5 \cdot 10^{-7}$ sec^{-1}	$1.13 \cdot 10^{-7}$ sec^{-1}
Photoionization rate ν_{He}	$0.23 \cdot 10^{-7}$ sec^{-1}	$0.52 \cdot 10^{-7}$ sec^{-1}
Solar wind proton flux f_p	$1.5 \cdot 10^8 \frac{\text{prot}}{\text{cm}^2\,\text{sec}}$	
Total Lyman-Alpha radiation intensity	$2.7 \cdot 10^{11} \frac{\text{phot}}{\text{cm}^2\,\text{sec}}$	$3.2 \cdot 10^{11} \frac{\text{phot}}{\text{cm}^2\,\text{sec}}$
Lyman-Alpha radiation pressure on hydrogen (K_s/K_g)	0.3	0.8

For the calculation of the typical maximum and minimum rate coefficients ν for photoionization we adopt the EUV-measurements of HINTEREGGER *et al.* (1963) to be representative for solar minimum conditions. With the use of these data we obtain the photoionization rates for solar minimum conditions:

$$\nu_{\mathrm{H,\,min}} = \int_0^{912\,\text{Å}} q_{i,\mathrm{H}}(\lambda) \cdot I_{\min}(\lambda) \cdot d\lambda = 0.5 \cdot 10^{-7}\,\mathrm{sec}^{-1} \tag{2}$$

for hydrogen and

$$\nu_{\mathrm{He,\,min}} = \int_0^{504\,\text{Å}} q_{i,\mathrm{He}}(\lambda) \cdot I_{\min}(\lambda) \cdot d\lambda = 0.23 \cdot 10^{-7}\,\mathrm{sec}^{-1} \tag{3}$$

for helium at $r = 1$ AU. Here $q_{i,\mathrm{H}}$ and $q_{i,\mathrm{He}}$ are the photoionization cross-sections of hydrogen and helium, and $I(\lambda)$ is the intensity of the EUV-radiation at the wavelength λ.

Typical maximum ionization rates may now be calculated from (2) and (3) by the use of the factor 2.25 for the maximum-to-minimum EUV-intensity ratio as given by a linear correlation fit of HALL *et al.* (1969). Therefore we may obtain the following solar-maximum rate coefficients:

$$\nu_{\mathrm{H,\,max}} = \int_0^{912\,\text{Å}} q_{i,\mathrm{H}}(\lambda) \cdot I_{\max}(\lambda) \cdot d\lambda = 1.13 \cdot 10^{-7}\,\mathrm{sec}^{-1} \tag{4}$$

and

$$\nu_{\mathrm{He,\,max}} = \int_0^{504\,\text{Å}} q_{i,\mathrm{He}}(\lambda) \cdot I_{\max}(\lambda) \cdot d\lambda = 0.52 \cdot 10^{-7}\,\mathrm{sec}^{-1} \tag{5}$$

at $r = 1$ AU.

As a representative solar cycle average of the solar wind proton flux we shall use $f_p = 1.5 \cdot 10^8$ prot/cm^2 sec at $r = 1$ AU. The charge exchange cross-section between hydrogen atoms and protons is taken to be $q_{ex} = 2 \cdot 10^{-15}$ cm^2 (FITE *et al.*, 1962). Summarizing all the results mentioned, we shall consider the following behaviour of active solar radiations as characteristic for a standard solar activity cycle.

III. The Solar Cycle Variation of the Heliospheric Cavity

Table 1 in chapter II gives typical maximum and minimum values for the solar radiations during the solar activity cycle. With these as typical ranges for the variation of linearly solar-activity-correlated radiations we can obtain a correlation of the mean penetration depth r_c of hydrogen with the sunspot number R_z. This correlation is shown in Fig. 2. The penetration depth r_c given here is the mean solar distance in upwind direction which can be reached by neutral interstellar hydrogen entering the solar system, before ionization occurs. The fully-drawn twin line shows that r_c due to the variations in the Lyman-Alpha radiation pressure and the EUV-radiation varies between 3 and 8 AU during a standard solar cycle. If we introduce the term "heliospheric cavity" for that circumsolar region which is not reached by the average interstellar neutral hydrogen atom, the mean penetration depth $r_c(\Theta)$ just defines the geometrical extent of this region of ionized

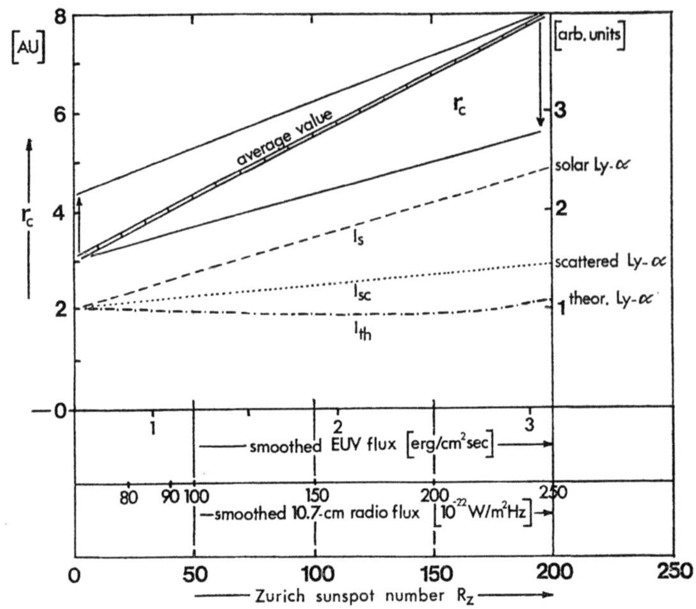

Fig. 2. The radius of the heliospheric cavity r_c (left ordinate) and the solar Lyman radiation I_s, the scattered extraterrestrial Lyman-Alpha-radiation I_{sc} and a theoretically-obtained value I_{th} (right ordinate in arbitrary units) are shown versus the smoothed Zurich sunspot number R_z, the smoothed 10.7 cm-solar radio flux, and the solar EUV-flux. The fully-drawn twin line represents r_c for a monthly average $\bar{R}_z = R_z$; the upper or lower full line show the possible deviations of r_c the actual value R_z is lower or higher than the monthly average $\bar{R}_z$

interstellar hydrogen. The extreme configurations between which the boundary $r_c(\Theta)$ of the heliospheric cavity varies during the standard solar activity cycle (see Table 1) is shown in Fig. 3. The inner configuration represents the solar minimum cavity, the outer configuration the solar maximum cavity. The geometrical shape of the cavity is nearly spherical in the upwind region, and is elongated in downwind directions into a more elliptical configuration which is strongly cusped at angles $160° < \theta < 200°$ opposite to the direction of approach of interstellar matter. This cusping is due to the enhanced densities of gravitationally focused hydrogen.

IV. Solar Cycle Behaviour of Scattered Lyman-Alpha Radiation

It might be interesting now to consider the Lyman-Alpha intensity I_{sc} which originates from the scattering of solar Lyman-Alpha radiation I_s at interplanetary hydrogen. (THOMAS and BOHLIN 1972) have given measurements of this intensity I_{sc} obtained with the OGO V satellite from a specific celestial direction over a period of about 2.5 years, from March 1968 to June 1971. A linear least-square fit of these observed intensities I_{sc} plotted versus sunspot number R_z shows a tendency of higher intensities I_{sc} to occur at higher activities R_z. This linear correlation given

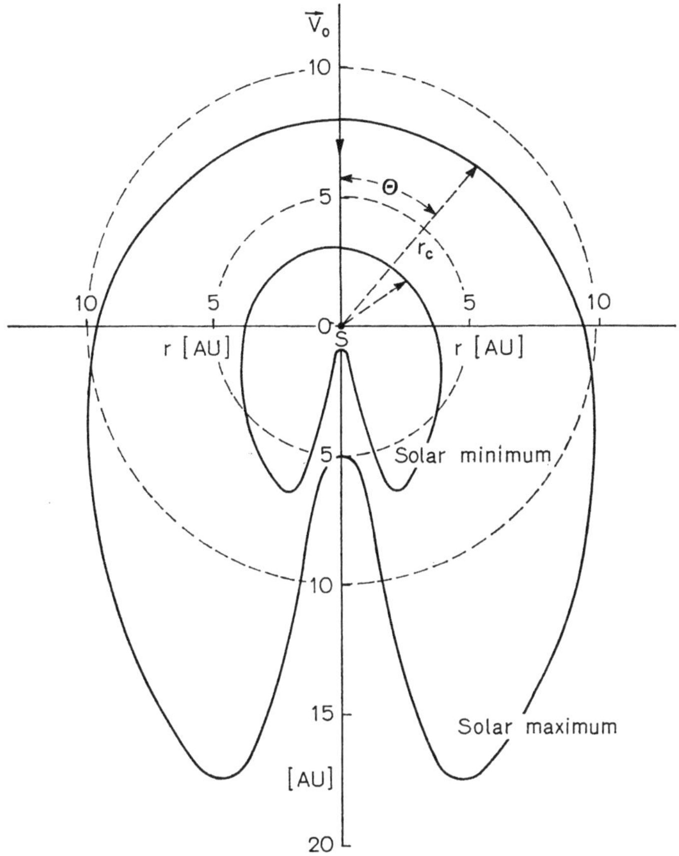

Fig. 3. The geometrical configuration of the heliospheric cavity $r_c(\theta)$ for solar minimum and solar maximum conditions. The average solar cycle variations of the ionizing solar fluxes are adopted as shown in table 1 of this paper. ($T=250$ K), $v_0=20$ km/sec, r means the solar distance; θ is the angle between $\boldsymbol{v}_0$ and $\boldsymbol{r}$

by the above mentioned authors is shown in Fig. 2. The intensity ratio $I_{sc}(\max)/I_{sc}(\min)$ turns out to be 1.44 in the range $0 \leqq R_z \leqq 200$ which is essentially lower than that of 2.4 for the solar Lyman-Alpha radiation I_s taken from BLAMONT and VIDAL-MADJAR. This fact points out that there must exist some counteracting effect in the physics of the scattered intensity I_{sc} which itself contains I_s as a multiplication factor. The counteraction comes from the reaction of the distribution of the scattering agent, namely atomic hydrogen, to varying numbers of R_z. The mean penetration depth r_c increases when R_z and I_s increase (see Fig. 2). It is therefore interesting to estimate roughly the theoretical dependence of I_{sc} on R_z and to compare it with the observationally detected trend of I_{sc}.

In upwind directions the density profile for the scattering agent, neutral interstellar hydrogen, can approximately be represented by the function:

$$n(r)=n_0 \exp(-r_c/r). \tag{6}$$

With this density distribution the theoretical intensity I_{th} of the scattered Lyman-Alpha radiation seen from the earth in upwind directions is given by:

$$I_{th} A_0 \int_{r_E}^{\infty} I_{s,E} (r_E/r)^2 \cdot \exp(-r_c/r)\, dr \tag{7}$$

where A_0 is a constant which contains the density n_0, the Einstein transition coefficient $B_{1,2}$ for the Lyman-Alpha transition and apparative parameters. $I_{s,E}$ is the solar Lyman-Alpha intensity at $r=r_E=1$ AU. With the introduction of $\rho=r_c/r$ and $\bar{r}_c=r_c/r_E$ we obtain from (7):

$$I_{th}=r_E \cdot A_0 \cdot I_{s,E}[1-\exp(-\bar{r}_c)]/\bar{r}_c. \tag{8}$$

Since r_c as shown in Fig. 2 varies between 3 and 8 AU, Eq. (8) may be simplified into the form:

$$I_{th}=r_E A_0 (I_{s,E}/\bar{r}_c). \tag{9}$$

In a theoretical first-order approach for the intensity I_{th} of the scattered Lyman-Alpha radiation we find I_{th} therefore to be inversely proportional to the penetration depth r_c. If we normalize I_{th} to 1 at $R_z=0$ and use the $r_c(R_z)$-relation given in Fig. 2, we obtain the curve I_{th} versus R_z given by the dashed-dotted line in Fig. 2. It is obvious that the curve for the observational trend of I_{sc} has systematic deviations from the curve I_{th}.

In order to give an interpretation for this systematic deviation, one has to consider the meaning and the physical nature of r_c as calculated in Fig. 2. Here r_c has been obtained as a function of R_z under the assumptions that for hydrogen prevail constant ionization conditions, which correspond to a constant activity R_z. This means that if a temporally constant R_z together with constant corresponding values for the solar EUV-flux and the solar Lyman-Alpha flux are given, a penetration depth $r_c(R_z)$ as given in Fig. 2 would characterize the upwind distribution of interstellar hydrogen in interplanetary space. Since, however, such solar conditions will never be fulfilled, $r_c(R_z)$ has to be interpreted as being that penetration depth which characterizes the distribution of hydrogen, if an average solar activity $\bar{R}_z=R_z$ prevails. This turns out to confer a significant meaning to r_c, since the hydrogen density distribution by its nature has the effect of smoothing-out and equalizing short-time variations in solar conditions, and does not reflect solar flux variations having periods shorter than a few months in r_c (FAHR and BLUM, 1971). This is because the ionization probability is mainly determined by the solar fluxes which prevailed during the last 1-Au-passage of a hydrogen atom along its orbit to a certain space point $\boldsymbol{r}$ in question, so that the actual value for r_c at a time t is related to the actual monthly average of $\bar{R}_z$ for the period that immediately preceded t.

This is the reason why the theoretical curve I_{th}, which is calculated with a penetration depth r_c belonging to an average activity $\bar{R}_z=R_z$, deviates from the observational intensity I_{sc}; I_{th} is a function containing a quantity I_s that varies instantaneously with varying solar activity R_z and a quantity $n_H(r, \theta)$ that only varies with the monthly average of R_z. Therefore a deviation of I_{sc} from I_{th} is explained by the fact that I_{sc} given by THOMAS and BOHLIN (1972) may be measured

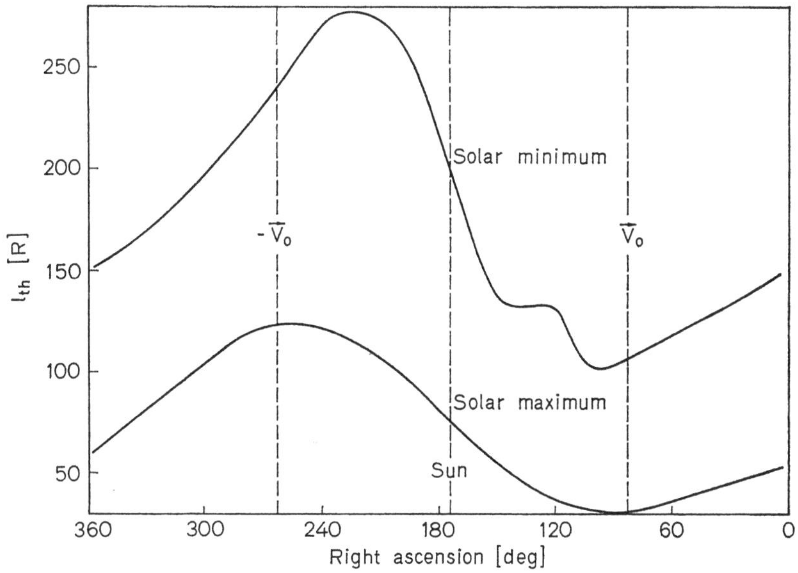

Fig. 4. The theoretically calculated Lyman-Alpha background radiation for solar minimum and maximum conditions (see Table 1) is given versus right ascensions of a celestial main-circle survey registrated from the position of the earth at Sept. 17. The plane of the main circle at the sky contains the position of the sun and the vector v_0 taken from the direction RA=265°, $\delta=-23°$ with an amount of 20 km/sec ($T=250$ K, $n_0=0.1$ cm^{-3}). The solar central line intensity I_s at 1216 Å has been taken to be 1.7 erg/cm^2 sec Å at solar minimum, and $2.4 \cdot 1.7=4.1$ erg/cm^2 sec Å at solar maximum

simultaneously with a solar activity R_z which differs greatly from the monthly activity average $\bar{R}_z$ that triggers the penetration depth r_c. For instance a value I_{sc} registrated simultaneously with a high activity level $R_z \approx 200$ cannot be compared with I_{th} calculated for $r_c=r_c(\bar{R}_z=200)$, if the actual average activity $\bar{R}_z$ is far below 200. In Fig. 2 the single full lines above and below the twin line showing $r_c(\bar{R}_z)$ give values for r_c which should be used in order to bring I_{th} into accordance with I_{sc}. In the upper full line we find the appropriate values r_c for an intensity I_{th} normalized to 1 at $R_z=200$, in the lower line appropriate values r_c for intensities I_{th} normalized to 1 at $R_z=0$. These normalizations are necessary since the function I_{th} contains some unknown parameters in the constant A_0 [see Eq. (7)], and therefore cannot be compared with the absolute value of I_{sc}. Since the monthly average $\bar{R}_z$ is always lower than R_z for high instantaneous activities $R_z \approx 200$, I_{sc} should be interpreted with the values for r_c given in the lower curve; whereas it should be interpreted with the values r_c given in the upper curve for very low instantaneous activities $R_z \approx 0$, since in that case $\bar{R}_z$ is certainly larger.

The intensity I_{sc} of scattered Lyman-Alpha radiation does not change only in one direction during the solar cycle; the total Lyman-Alpha sky background seen from a fixed point in the interplanetary space undergoes a systematic solar-cycle-correlated change. This can be seen in Fig. 4, where a theoretical scan of the scattered Lyman-Alpha intensities I_{th} along the v_0-axis seen from the position of the earth on Sept. 19 both for solar maximum and solar minimum conditions (see Table 1) is presented. This scan is part of a celestial main-circle survey in the plane containing the sun and the vector $\boldsymbol{v}_0$ of the velocity of approach of interstellar matter. In the isophotes of the Lyman-Alpha background radiation taken with the UV-spectrometers of the OGO V satellite BERTAUX and BLAMONT (1971) and THOMAS and KRASSA (1971) have found an indication for the vector $(-\boldsymbol{v}_0)$ to point into the celestial direction $\alpha=265°$, $\delta=-23°$ (α-right ascension, δ-declination). We shall also adopt this position of the vector $\boldsymbol{v}_0$. The comparison of the two main-circle scans for solar maximum and solar minimum conditions reveals important qualitative and quantitative differences which are mainly due to the solar cycle changes of the heliospheric cavity shown in Fig. 3. Whereas the main contribution to the scattered intensity comes from regions outside the orbit of the earth during solar maximum conditions, they come from regions near the earth's orbit during solar minimum conditions. The Lyman-Alpha sky background for minimum conditions resembles qualitatively the interplanetary Helium-I-584 Å sky background calculated by FAHR and LAY (1972), since the main contributions to the 584 Å radiation intensity come also from regions inside the earth's orbit, this being the case, however, for the whole solar activity cycle.

V. Solar Wind Deceleration and the Solar Cycle Variation of the Heliopause

Solutions derived from hydrodynamical treatments of the solar wind expansion (PARKER, 1962, 1963; HARTLE and BARNES, 1969; HARTLE and STURROK, 1968) without taking account of the interaction of solar wind plasma with the interstellar medium yield a constant supersonic expansion velocity v_{sw} beyond a typical solar distance of $r=0.5$ AU. For such non-interacting models it is easy to calculate a distance r_h where, due to the geometrical divergence of the plasma flow, the solar wind kinetic energy-density has decreased to the energy-density $(B_i^2/8\pi)$ of the interstellar magnetic field. Beyond r_h the expansion of the solar wind is controlled by the magnetic field and cannot continue to go in radial directions. This is the reason why we would like to call the geometrical boundary of the heliosphere defined by the solar distance r_h the "heliopause".

The location of the heliopause is strongly affected by the interaction of the solar wind with the interstellar medium. Interstellar hydrogen and helium atoms which enter the solar system with a velocity $\boldsymbol{v}_0$ become ionized while approaching the sun. These ionization processes lead to the generation of secondary helium and hydrogen ions in the solar wind which are re-accelerated by the frozen-in magnetic fields of the solar wind plasma to the bulk speed v_{sw} of the solar wind. This re-acceleration process is connected with a mean loss of solar wind kinetic energy per secondary ion (FAHR, 1971c, d) given by:

$$E_{kin}=-1/2\, m_i v_{sw}^2 (1+2\sin^2\alpha-\sin^4\alpha) \tag{10}$$

where α gives the angle between $\boldsymbol{v}_{sw}$ and the frozen-in magnetic field $\boldsymbol{B}_f$, and m_i is the mass of the secondary ion. This description in Eq. (10) takes account of the effects of momentum loss due to the transcharging of a solar wind proton, of the momentum gain due to the re-acceleration of the generated secondary ion, and of the momentum loss due to the gain of suprathermal energies by the helicoidally gyrating secondary ion, taken from the linear momentum density $n_p m_p v_{sw}$ of the solar wind plasma. This description must be expected to be sufficiently correct as long as the kinetic energy $n_p m_p v_{sw}^2$ of the solar wind is dominant over the thermal and magnetic energies $n_p \mathrm{kT}_p$ and $B_f^2/8\pi$ of the plasma. Since it is not yet clear whether the distant solar wind is weakly or strongly shocked, or even shocked at all (WALLIS, 1972), and for which parameters it will undergo a transition from supersonic to subsonic expansion velocities in the outer heliospheric regions (HOLZER, 1972), we shall tacitly assume here these conditions to be fulfilled in the region which is of interest in the following calculations.

Other descriptions of the interaction of the interstellar medium with a radially expanding magnetized solar wind plasma have been tried by HOLZER (1972) and WALLIS (1971), but have also been found to be not yet satisfactory in a paper of WALLIS (1972) re-examining the theoretical approaches to this problem. The main insufficiencies in these descriptions seem to be that HOLZER (1972) tries to discuss the effect of the frozen-in magnetic field B_f by including a Lorentz-force term into the momentum equation and an Ohm's heating term into the energy conservation equation of his hydrodynamical one-fluid model via Maxwell's equations, but uses an inconsistent $(r v_{sw})^{-1}$-solution for B_f. WALLIS (1971) formally includes the field B_f via the total pressure $(P+B_f^2/8\pi)$ in the momentum equation; however, he neglects its contribution to the energy flux. Neither HOLZER, nor WALLIS adequately take into account the effect producing suprathermal energies due to the helicoidal motions of secondary ions around the streaming field lines of B_f. These energies are primarily taken from the frozen-in field, but finally from the linear momentum of the plasma which transports the field. Therefore this production of suprathermal energies results finally in a loss of linear momentum and leads to an additional coupling between the equation of momentum and the equation of energy conservation. This coupling has not been taken into account in the above-mentioned treatments.

We shall therefore follow a treatment published by FAHR (1971d). Since the energy and momentum loss due to interacting interstellar particles starts to become important for the solar wind dynamics outside of $r=5$ AU, we can take the angle α in Eq. (10) to be $\pi/2$. In that case the following differential relation for the exhaustion of kinetic energy from the solar wind is valid:

$$dE_{\mathrm{kin}}=d(m_p v_{sw}^2/2)=dr\,\frac{v_{\mathrm{rel}}}{v_{sw}]}\,[q_{ex}(p,\mathrm{H})\cdot \Delta E_{\mathrm{H}}\cdot n_{\mathrm{H}}+q_{ex}(p,\mathrm{He})\cdot \Delta E_{\mathrm{He}}\cdot n_{\mathrm{He}}] \tag{11}$$

where v_{rel} is the relative velocity between the solar wind protons and the neutral species, $q_{ex}(p, \mathrm{H})$ and $q_{ex}(p, \mathrm{He})$ are the charge exchange cross-sections for proton-hydrogen and proton-helium collisions, and ΔE_{H} and ΔE_{He} are the specific energy losses dE_{kin} per secondary proton and helium ion given in Eq. (10). Eq. (11) can be integrated to yield:

$$\begin{aligned} n_p(r,\theta)&=n_{p,E}\left(\frac{r_E}{r}\right)^2\cdot W(r,\theta)^{-1}\\ v_{sw}(r,\theta)&=v_{sw,E}\cdot W(r,\theta)\end{aligned} \tag{12}$$

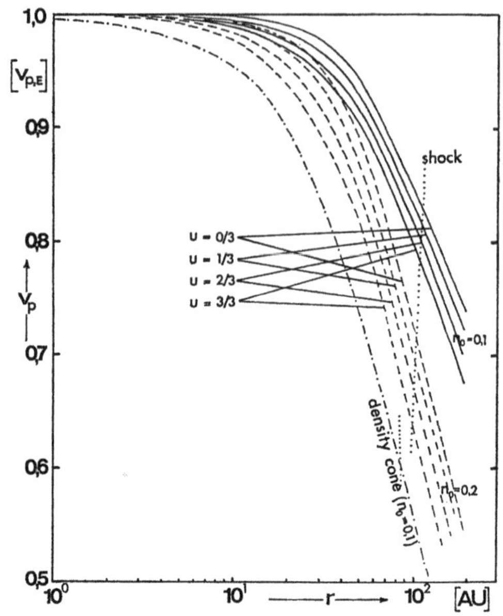

Fig. 5. The solar wind velocity v_p is shown versus the distance r from the sun. The dotted line in the nearly vertical direction gives the location of the heliospheric shock front. The parameter $\mu=(K_g-K_s)/K_g$ gives the net gravitational force for different solar Lyman-Alpha pressures

where the function W for the extraction of kinetic energy from the solar wind is given by:

$$W(r,\theta)=\exp\left[-q_{\mathrm{ex}}(p,\mathrm{H})\int_{r_0}^{r} n_{\mathrm{H}}(r,\theta)\,dr-q_{\mathrm{ex}}(p,\mathrm{He})\int_{r_0}^{r} 4\,n_{\mathrm{He}}(r,\theta)\,dr\right]. \quad (13)$$

With the use of the densities of interstellar helium and hydrogen in interplanetary space shown in Fig. 1 we can now calculate the deceleration of the solar wind plasma flow. The result is shown in Fig. 5 for upwind and downwind density profiles belonging to different solar Lyman-Alpha radiation pressures. One can see that the deceleration of the solar wind becomes important beyond $r=10$ AU. Especially pronounced is the deceleration in the downwind regions, where the enhanced densities due to gravitational focusing cause enhanced extraction of linear momentum from the downwind solar wind plasma flow.

From the radial profile of the kinetic energy density $n_p\,m_p\,v_{\mathrm{sw}}/2$ given by Eq. (12) we can easily find the location r_h of the heliopause by simply determining the distance $r_h(r,\theta)$ where this energy density equals that of the interstellar field B_i. In Fig. 6 the resulting configuration for the heliopause is shown for solar maximum and solar minimum conditions (see Table 1). The solar cycle variation of the geometrical shape of the heliopause is found to be rather unimportant. Only in the downwind region can a slight solar cycle variation of r_h be expected.

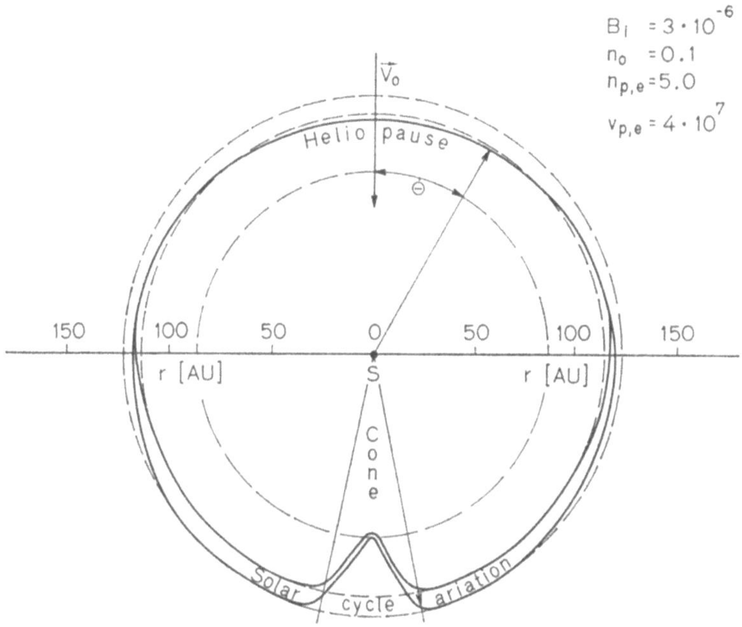

Fig. 6. Shown is the location of the heliopause $r_h(\theta)$ for solar minimum and solar maximum conditions (see Table 1). The polar co-ordinates are the solar distance r and the angle θ with respect to the upwind direction. The cusped region is due to the enhanced energy extraction from the solar wind by gravitationally focused hydrogen and helium of $T=1000$ K. ($v_0=20$ km/sec)

VI. Conclusions

In Fig. 1 to 6 we have presented solar-cycle-correlated variations of heliospheric properties like the heliospheric densities n_{H} and n_{He} of neutral interstellar helium and hydrogen, the distance r_c of the heliospheric cavity, the intensity I_{th} of the Lyman-Alpha background radiation, the dynamical parameters v_{sw} and n_p of the distant solar wind, and the distance r_h of the heliopause. Due to the fact that, different to the solar wind proton flux f_p, the solar Lyman-Alpha flux I_s and the solar EUV-flux f_{EUV} are strongly activity-correlated, all the above-mentioned quantities vary more or less significantly during the average solar cycle.

Especially pronounced is this variation with respect to the shape of the heliospheric cavity. This means the neutral interstellar medium is excluded from a circumsolar volume which is much larger at solar maximum than at solar minimum. This difference in the density distribution of interstellar hydrogen in the interplanetary space also affects the isophotes of the extraterrestrial Lyman-Alpha background radiation as seen from the earth. As can be seen in Fig. 4 the solar maximum radiation pattern differs appreciably both in quantity and quality from the minimum radiation pattern. This could be observationally confirmed when measurements done by Bertaux and Blamont (1971) and Thomas and Krassa (1971) with the OGO V satellite during 1969/70 will be repeated during

1973/74. The solar cycle effect on the dynamics of the solar wind at distant solar regions ($r > 10$ AU) does not seem to be very exciting. The small influence of the activity-correlated density profiles is due to the fact that these profiles only change at regions inside of $r \approx 5$ AU, whereas the energy extraction from the solar wind as discussed in chapter V becomes important only beyond $r \approx 10$ AU. Due to the same reason the variations Δr_h in the distance r_h of the heliopause also remain unimportant during the solar cycle.

References

AXFORD, W. I.: Asilomar Solar Wind Conference California, March 1971.
BERTAUX, J. L., BLAMONT, J. E.: Astron. Astrophys. **11**, 205 (1971).
BLAMONT, J. E., VIDAL-MADJAR, A.: J. Geophys. Res. **76**, 4311 (1971).
BLUM, P. W., FAHR, H. J.: Astron. Astrophys. **4**, 280 (1970).
FAHR, H. J.: Astron. Astrophys. **14**, 263 (1971a).
FAHR, H. J.: Planet. Space Science **19**, 1121 (1971b).
FAHR, H. J.: Preprint 71-10 Astronomische Institut Bonn (1971c).
FAHR, H. J.: Cosmic Plasma Physics Conference ESRO Frascati (Rome) Sept. 1971d.
FAHR, H. J., BLUM, P. W.: Astron. Astrophys. **8**, 226 (1971).
FAHR, H. J., LAY, G.: Preprint COSPAR Symposium 1972 Madrid, June 1972.
FELDMAN, W. F., LANGE, J. J., SCHERB, F.: Asilomar Solar Wind Conference California, March 1971.
FITE, W. L., SMITH, C. H., STEBBINGS, R. F.: Proc. Royal Society A-268, 527 (1962).
HALL, L. A., HIGGINS, J. E., CHAGNON, C. W., HINTEREGGER, H. E.: J. Geophys. Res. **74**, 4181 (1969).
HARRIS, I., PRIESTER, W.: NASA Special Report D/1444 (1962).
HARTLE, R. E., BARNES, A.: J. Geophys. Res. **75**, 6915 (1969).
HARTLE, R. E., STURROK, P. A.: Astrophys. Journal **151**, 1155 (1968).
HINTEREGGER, H. E., HALL, L. A., SCHMIDTKE, G.: Space Research V, 1175 (1963).
HIRSHBERG, J., ASBRIDGE, J. R., ROBBINS, D. E.: J. Geophys. Res. **77**, 3583 (1972).
HOLZER, T. E.: Preprint Boulder (Colorado) 80302 (1972).
HOLZER, T. E., AXFORD, W. I.: J. Geophys. Res. **76**, 6965 (1971).
HUNDHAUSEN, A. J.: Reviews of Geophysics and Space Physics **8**, 729 (1970).
MIHALOV, J. D., WOLFE, J. H.: Cosmic Electrodyn. **2**, 326 (1971).
PARKER, E. N.: Planet. Space Science **9**, 461 (1962).
PARKER, E. N.: In: Interplanetary dynamical processes. New York: Interscience 1963.
ROBBINS, D. E., HUNDHAUSEN, A. J., BAME, S. J.: J. Geophys. Res. **75**, 1178 (1970).
SLUTZ, R. J., GRAY, T. B., WEST, M. L., STEWART, F. G., LEFTIN, M.: Preprint, Boulder (Colorado) November 1971.
THOMAS, G. E.: Asilomar Conference Report California, March 1971.
THOMAS, G. E., BOHLIN, R. C.: J. Geophys. Res. **77**, 2752 (1972).
THOMAS, G. E., KRASSA, R. F.: Astron. Astrophys. **11**, 218 (1971).
TOUSEY, R.: Space Science Rewies **2**, 3 (1963).
WALLIS, M.: Nature, Phys. Sciences, **233**, 23 (1971).
WALLIS, M.: Preprint Trita-Epp-71-01 Feb. 1971.
WALLIS, M.: Preprint Trita-Epp-72-03 June 1972.

Evidence of a Double Peak in Solar Maxima

By MICHAEL D. PAPAGIANNIS[1]
Dept. of Astronomy, University of Athens, Athens, Greece

CHRISTOS S. ZEREFOS and CHRISTOS C. REPARIS
National Hellenic Research Foundation, Athens, Greece

With 6 Figures

Abstract

A latitudinal analysis of solar flares accompanied by type-IV radio bursts during the maxima of cycles 19 and 20 indicates that solar maxima for the more energetic events consist of two independent peaks, the first of which is stronger at higher latitudes and the second at lower latitudes.

A list of nearly 350 flares accompanied by type-IV radio bursts by KRÜGER *et al.* (1971), which covers a period of 14 years (1956–1969), was expanded to include all SID, PCA and Solar Cosmic Ray events during the above mentioned period. Type-IV flares are well correlated with the more energetic events of the sun (KUNDU, 1965). This is evident also from the fact that with our 350 type-IV flares were associated 287 SID events, 144 PCA events, 105 Solar Cosmic Ray events and all of the 13 Ground Level events which occurred during the period 1956–1969.

This large sample of the more energetic solar events was used to examine their time and latitude distribution during each solar cycle. Our analysis has shown that all these energetic events display a double maximum in both of the two solar cycles (19 and 20) we studied. This confirms the basic conclusion which GNEVYSHEV and his collaborators (1963, 1965, 1966, 1967) had reached without the data for cycle 20. KRIVSKY and KRÜGER (1966), on the other hand, had investigated the period 1954–1965 and found a second maximum in 1960 in the total number of flares per year which were accompanied by type-IV radio bursts. Unfortunately there are no reliable radio data available for the maximum of cycle 18 because at that time solar radio Astronomy was just starting.

It should be noted here that in a statistical study of this nature one must use periods considerably shorter than one year, because the time lag between the two maxima is of the order of two years. Hence, the grouping of different events of solar activity under study in 12-month periods, might strongly reduce or even cancel the distinctiveness of the two peaks. This is clearly seen in Fig. 1 which show the distribution of type-IV flares in monthly, semi-yearly and yearly intervals. It is clear that the double peak in the maximum of cycle 20, which is quite prominent in the monthly and semi-yearly groupings of events, disappears when the events are grouped in yearly intervals.

Fig. 2a shows the total number of type-IV flares and the number of PCA events associated with them (shaded areas) in each 6-month interval during the period 1956–1969. Fig. 2b gives the same data, except that only one event per

1 On Sabbatical leave from the Dept. of Astronomy, Boston University, Boston, Mass. U.S.A.

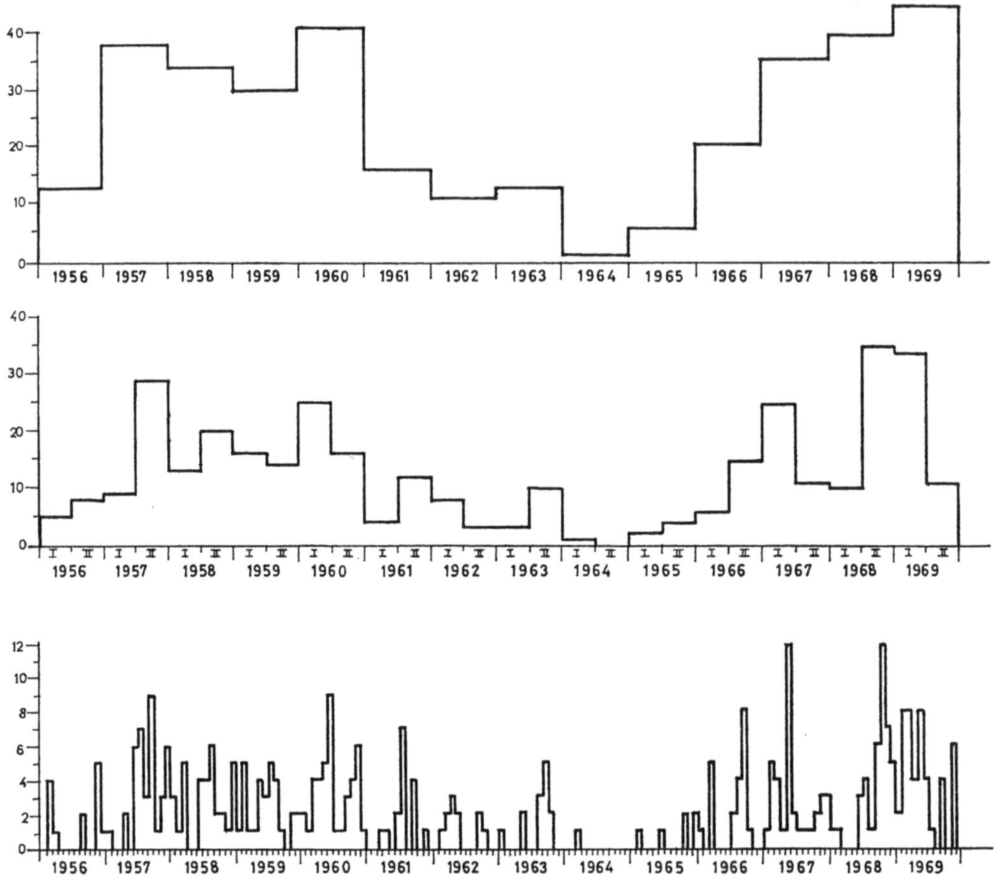

Fig. 1. The total number of type-IV flares in yearly (upper histogram), semi-yearly (middle histogram) and monthly (lower histogram) intervals during the period 1956–1969, which shows the disappearance of the two-peak structure in the period 1967–1969 in the yearly grouping of the events

plage region is included. The appearance of the two maxima is still quite evident, but from a comparison of the two figures it follows that the average number of type-IV flares per plage region is considerably higher in the second peak of both cycles. It is also obvious that we have more PCA events, as well as more PCA events per plage region in the second peak of these cycles.

Fig. 3 shows the total number of proton events associated with type-IV radio bursts as well as the number of Ground Level Events (open circles) during the period 1956–1969. The shaded areas give the number of proton events counting only one event per plage region, or alternatively the number of plage regions in which these events occured. Fig. 3 shows again clearly the appearance of the two peaks in the maxima of both cycles under consideration and shows also that the second maximum includes a considerably higher number of proton flares and GLE's.

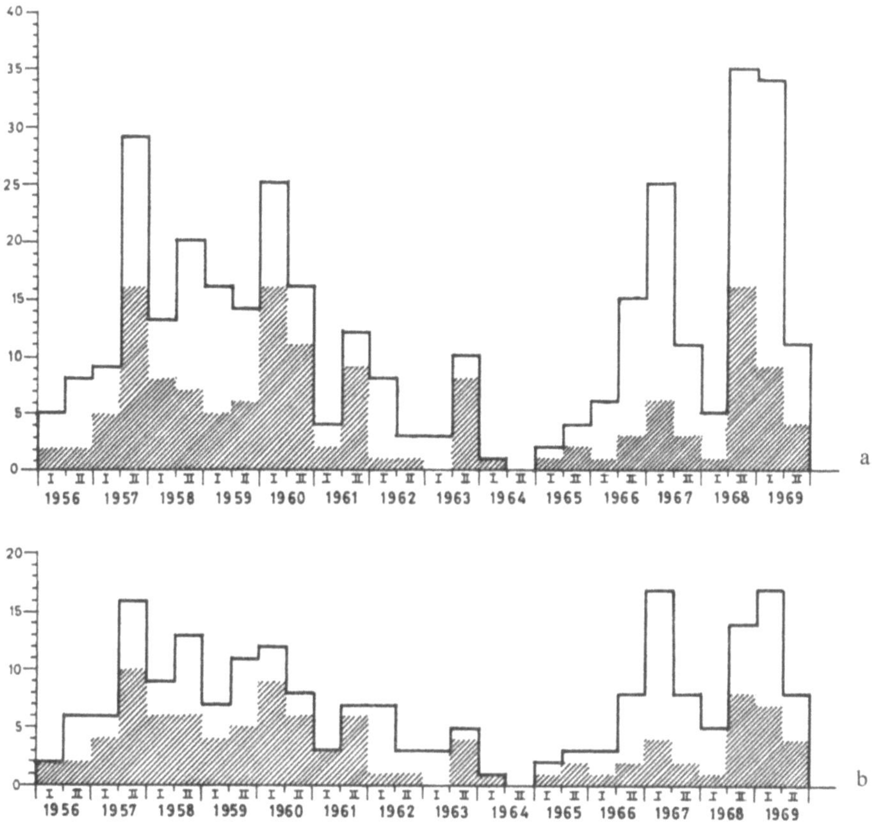

Fig. 2. The total number of type-IV flares and those associated with PCA events (shaded areas) in each 6-month interval during the period 1956–1969 (a). The same data taking, however, only one event per plage region are shown in (b)

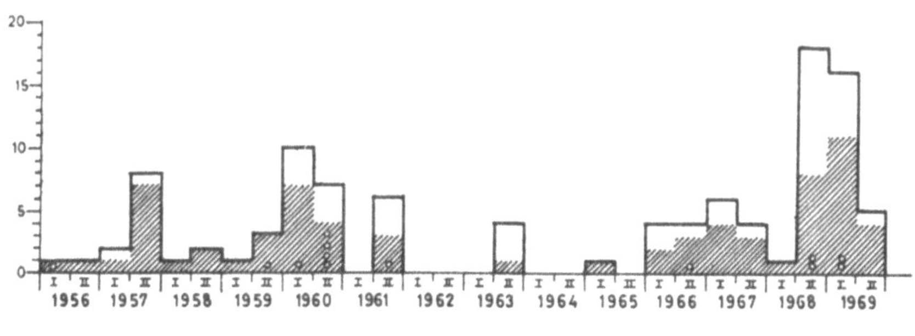

Fig. 3. The total number of proton events (associated with type-IV bursts) in each 6-month interval from 1956 to 1969. The shaded areas give the number of plage regions in which these events were produced, and the open circles the Ground Level Events which occured during this period

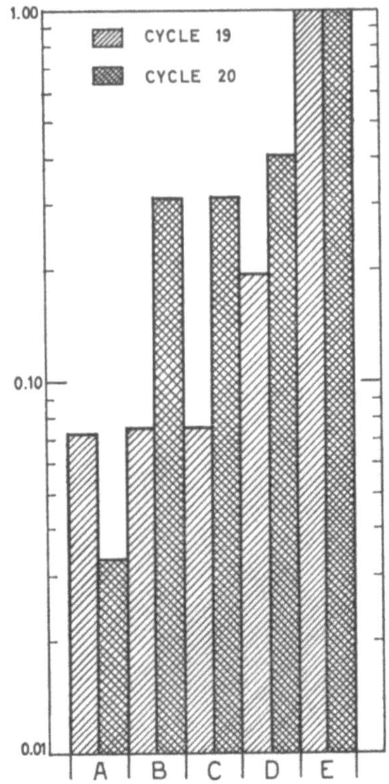

Fig. 4. A diagram showing the increase of the strength of the second peak relative to the first, i.e. $(X_2-X_1)/(X_2+X_1)$, as we move toward the more energetic events ($X=A$, B, C, D, E) in both cycles 19 and 20

An important conclusion which can be drawn from the above analysis is that as we move toward the more energetic events in a hierarchy of solar activity (type-IV flares, SID's, plage regions producing type-IV flares, proton events and GLE's), we have an intensification of the two-peak structure of the solar maximum and a strengthening of the second peak relative to the first. This is demonstrated in Fig. 4 where using as reference point the number of type-IV flares, we have computed the following percentages:

A = SID's per Type-IV flare
B = Plage regions per Type-IV flare
C = PCA's per Type-IV flare
D = Proton Flares per Type-IV flare
E = GLE's per Type-IV flare

for all four 6-month intervals which correspond to the first and to the second peak of each one of the two cycles. We have then used these percentages to obtain the ratio,

$$\frac{X_2-X_1}{X_2+X_1}$$

where X_1 and X_2 represent respectively the values of A, B, C, D and E at the first (1957-II or 1967-I) and the second (1960-I or 1968-II) peak of each cycle, in order to show that the predominence of the second maximum increases as we move to the more energetic events.

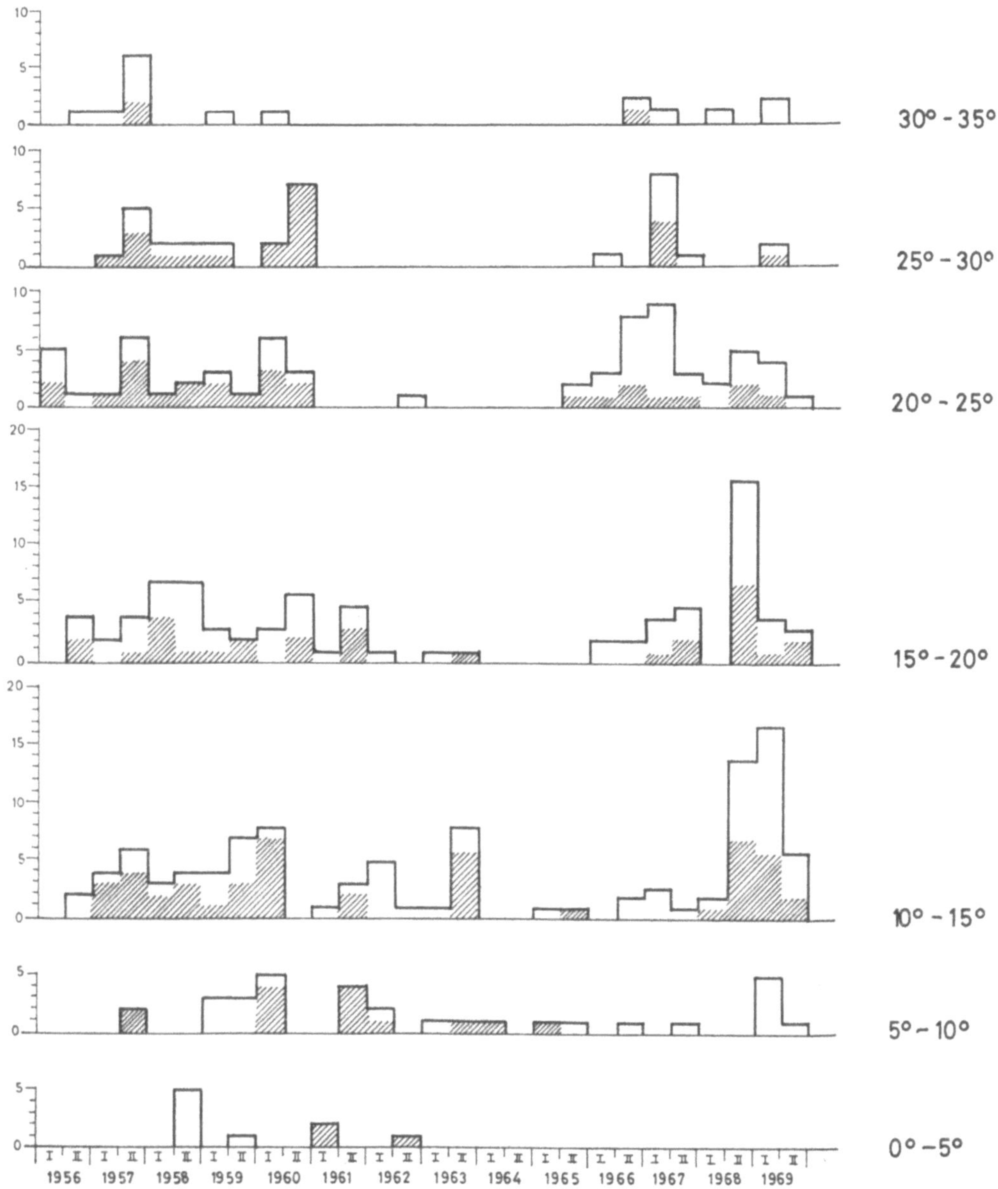

Fig. 5. The total number of type-IV flares and those associated with PCA events (shaded areas) in each 6-month period from 1956 to 1969 as distributed in seven 5° heliographic latitude zones (North and South taken together)

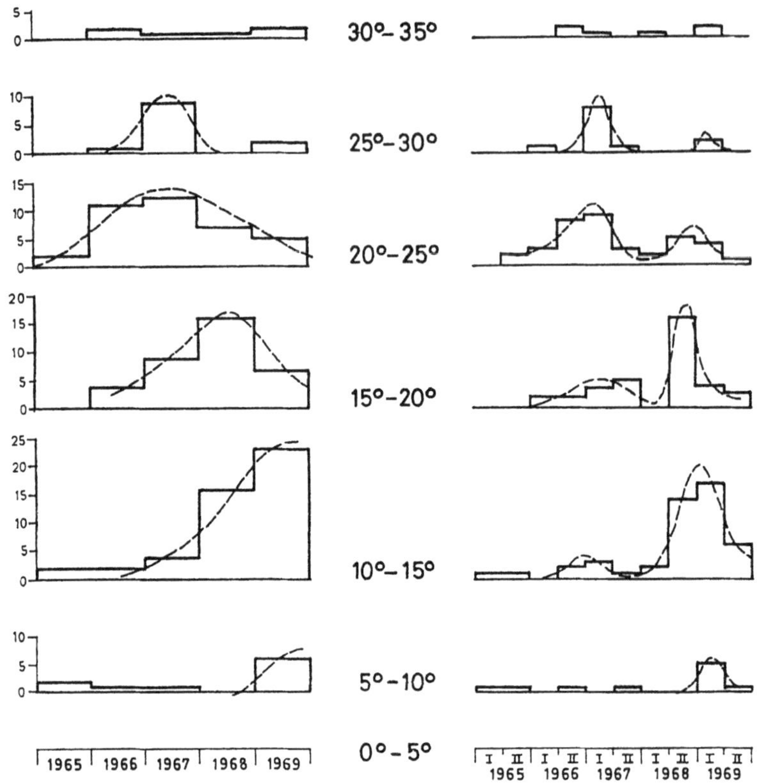

Fig. 6. The latitudinal distribution of the type-IV flares in yearly and semi-yearly intervals during the maximum of cycle 20, which show how the two-peak structure which is clearly evident in the 6-month grouping can be interpreted as a single broad peak drifting with time toward lower latitudes in the one-year grouping of the events

In addition to the above mentioned statistical studies we have performed also a latitudinal analysis of our data in order to investigate GNEVYSHEV'S suggestion that the two maxima in each solar cycle are independent, which he had based on his studies of the green coronal line during cycle 19. The distribution of all the type-IV flares in 5° latitude zones (North and South flares taken together) is shown in Fig. 5. This figure includes also (shaded areas) the latitudinal distribution of the PCA events associated with the type-IV flares.

The analysis for cycle 20 shows clearly that the two peaks represent two separate effects. The first peak has its highest strength in the 20–30° zone of heliographic latitude, while the second peak, which occurs approximately 2 years later, reaches its maximum strength in the 10–20° latitude zone. It is important to note also that the high-latitude peak does not shift gradually toward lower latitudes, as one might have expected from SPÖRER'S law (the butterfly pattern). Thus in the interval between the times at which the high latitude and the low latitude peaks occur we simply have a minimum at all latitudes. Also at the intermediate latitudes during the span of a solar cycle we seen the development of two separate peaks rather than the expected appearance of a single peak.

As it was mentioned earlier, however, the two peaks combine to give a single broader peak when we use a long enough (e.g. one year) grouping period. In this case, as seen in Fig. 6, we will observe a gradual drifting of this broader peak toward the lower latitudes as the cycle advances. This, however, is not a true effect, but simply the result of a coarser analysis of the data and the fact that the relative weight of the two peaks shifts gradually from the first to the second as we go from the higher to the lower latitudes.

The latitudinal analysis for cycle 19 did not produce as clear a distinction of the two peaks as it did for cycle 20. Instead one can see several peaks and there is no strong evidence of a systematic changing of the relative strength of the peaks with latitude. The difference of the 19th cycle from the 20th can possibly be attributed to the much higher level of sunspot activity which occurred during the 19th cycle and which probably tended to produce active centers over a much wider range of latitudes during the entire duration (1957–1960) of this solar maximum. Nevertheless when we consider all the events together without regard to latitude, this cycle, as we have seen before, shows also a double peak structure which becomes more pronounced as we move toward the more energetic events.

Acknowledgements

This research was supported in part by Air Force contract No. 203 F19628-71-0027 to Boston University, and in part by the National Hellenic Research Foundation, to the President of which, Mr. A. Natsinas, the authors express their appreciation.

References

Gnevyshev, M. N.: Soviet. Astron. AJ. **7**, 311 (1963).
Gnevyshev, M. N.: Soviet. Astron. AJ. **9**, 387 (1965).
Gnevyshev, M. N.: Solar Phys. **1**, 107 (1967).
Gnevyshev, M. N., Krivsky, L.: Soviet. Astron. AJ. **10**, 304 (1966).
Krivsky, L., Krüger, A.: Bull. Astron. Inst. Czech. **17**, 243 (1966).
Krüger, A., Fürstenberg, F., Friche, K. H.: 1971, HHI Suppl. Ser. Solar Data, Berlin II, 187 (1971).
Kundu, M. R.: Solar radio astronomy. New York, N. Y.: J. Wiley & Sons 1965.

Discussion

Xanthakis, J.:
The flare frequency curve presents one peak while the frequency curve of proton flares presents three peaks, one during the second quarter of 1957, one during the first quarter of 1960 and one in 1963 arround minimum activity.

Zerefos, Ch. S.:
The proton flare curve with IV Type Radioemissions which we have investigated does not present any appreciable peak for 1963 because during the 1957 peak we had eight type IV proton flares, during the 1960 peak we had eleven Type IV proton flares, while during the second semester of 1963 we had only four proton flares.

Besides according to GNEVYSHEV and KRISKY (Soviet Astronomy, 1966) the total number of proton flares were:

1957: 15 proton flares
1960: 18 proton flares
1963: 4 proton flares

MARIS, G.:
Have you investigated the double peak for North and South, separately?

ZEREFOS, CH. S.:
North and South bands of the same latitude were counted together to neutralize a rather pronounced North-South assymetry which occurs in our data. This was noted also by WALDMEIER (1971, Solar Physics) who concluded that the decade 1959–1969 was a period of strong *N–S* assymetry.

VALNICEK, B.:
A few years ago the same phenomenon was studied by GNEVYSHEV on the solar-corona and by G. VISNGVCOVÀ. The phenomenon was seriously documented by both authors.

ZEREFOS, CH. S.:
The basic purpose of our work was to study the variation of the number of flares associated with type-IV radio bursts at different latitude zones during the maxima of two consecutive solar cycles, and to investigate for cycle 20 the results which several authors had obtained for cycle 19 (see also references in that paper and another by PAPAGIANNIS *et al.* (1972) Solar Physics (in press).

Characteristic Features of Secular and Supersecular Cycles of Solar Activity. A 180-Year Cycle

By ANGEL D. BONOV
Department of Astronomy, Faculty of Physics, University of Sofia, Bulgaria

With 2 Figures

Abstract

Both 11-year cycles, the even and the odd ones according to the Zürich numeration, form a whole, i.e. the 22-year cycle of the solar activity. Therefore they cannot be separated the one from the other. This is why the secular and the super-secular cycles of the solar activity must necessarily begin with an even 11-year and must end with an odd 11-year cycle.

The author has found out a 180-year cycle. On this basis he draws the conclusion that until the end of the 20th century the solar activity will decrease considerably in comparison with its level at the 22-year cycle (18–19).

Cyclic Changes

Cyclic change is characteristic of solar activity. The basic cycle is the 11-year cycle.

The law discovered by HALE and NICHOLSON (1925) concerning the change of magnetic polarity of the bipolar groups at the beginning of each 11-year cycle proved the existence of the 22-year cycle.

GNEVYSHEV and OL (1948) pointed out that the even and odd 11-year cycles (according to the Zürich numeration) are closely bound together and form a whole. Hence they cannot be separated; to separate them would signify to disregard the existence of the 22-year cycle. This was made evident also by BONOV and CHRISTIAKOV (1959).

Taking into consideration the existence of the 22-year cycle we believe that the boundaries between the consistant secular cycles (secular minima) cannot be between an even and an odd 11-year cycle. The secular cycle may cover a different number of 22-year cycles, but it must necessarily contain an even number of 11-year cycles. We draw the same conclusion also for the supersecular cycle, whose duration is longer than that of the secular cycle.

The secular and the supersecular cycles must begin with an 11-year cycle, having an even number, and they must end with an 11-year cycle, having an odd number. Only under this condition the entirety of the 22-year cycle is preserved.

On the basis of the observations of solar spots OPENHEIM (1925), ANDERSON (1954) *et al.* draw the conclusion that there is a supersecular cycle.

Taking into consideration that the growth (T) of the 11-year cycle is a parameter with which the basic characteristics of the cycle are closely connected

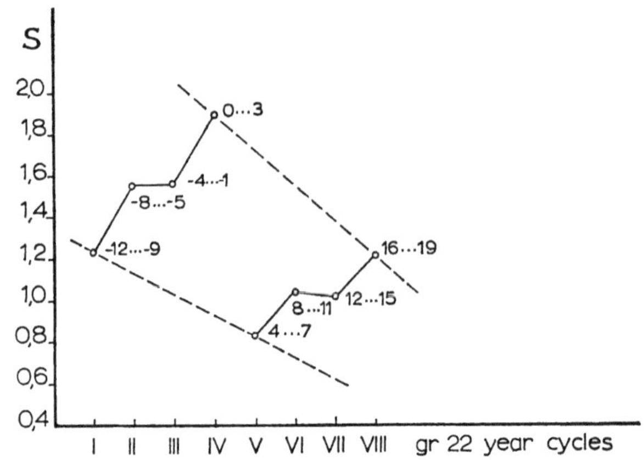

Fig. 1. The "180-year" cycle of the solar activity

Table 1

A group of 22-year cycles	22-year cycles	$T_{2n}+T_{2n+I}$	S
I	–12, –11	11.7	1.23
	–10, – 9	9.5	
II	– 8, – 7	10.0	1.56
	– 6, – 5	9.0	
III	– 4, – 3	11.3	1.57
	– 2, – 1	8.7	
IV	0, 1	11.6	1.90
	2, 3	6.1	
V	4, 5	10.3	0.83
	6, 7	12.4	
VI	8, 9	7.9	1.05
	10, 11	7.5	
VII	12, 13	9.5	1.02
	14, 15	9.3	
VIII	16, 17	8.4	1.22
	18, 19	6.9	

according to WALDMEIER (1935), we find the sum $T_{2n}+T_{2n+I}$ for all 22-year cycles after the year of 1610: (–12, –11), (–10, –9) ... (0–1), (2–3), ... (16–17), (18–19). Grouping them two by two, as it is shown on Table 1, we calculate

$$S=\frac{(T_{2n}+T_{2n+I})'}{(T_{2n}+T_{2n+I})''}.$$

The change of S is shown on Fig. 1. We see that after 1610 the solar activity, characterized by the index S, is presented by two supersecular cycles:

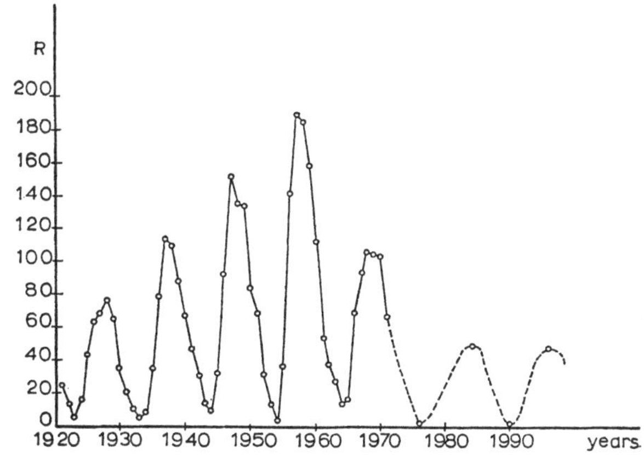

Fig. 2. A prognosis for the solar activity until the end of the 20th century

Table 2

No. of the 11-year cycle	Epoch of the min. (beginning)	Epoch of the max.	R_{max}	Epoch of min. (end)
20	1964,7	1968,4	106	1976,5 ± 0.4
21	1976,5 ± 0.4	1984,1 ± 0.5	50 ± 10	1990,6 ± 0.6
22	1990,6 ± 0.6	1996,6 ± 0.6	48 ± 12	

The first covers eight 22-year cycles. It begins in 1610,8 and ends in 1784,7. Its duration is 174 years.

The second covers also eight-year cycles. It begins in 1784,7 and ends in 1964,7. Its duration is 180 years.

Since the duration of the 22-year cycle is not constant, the duration of the supersecular cycle (BONOV, 1957), found by us, cannot be constant too. We call it conditionally a "180-year cycle".

GNEVYSHEV and OL (1948) mention that at the 22-year cycle (4–5) there is a "catastrophical" decrease of solar activity. According to our opinion this is due to the transitional epoch in the supersecular change of the solar activity. This can be seen on Fig. 1. Taking into consideration that with the 22-year cycle (20–21) a new supersecular cycle (Fig. 1) begins, we draw the conclusion that at present a "catastrophical" decrease of solar activity will take place anew. This conclusion is corroborated by comparing both 11-year cycles No. 19 and No. 20. Actually the 22-year cycle (20–21) will resemble the 22-year cycle (4–5). Up to the end of the century we live in the solar activity will be considerably weaker in comparison with its level at the 22-year cycle (18–19). This can be seen on the graph of Fig. 2, constructed on the basis of Table 2.

We obtained the elements in Table 2 on making use of the correlations found by us between the basic characteristics of the 11-year and the 22-year cycles (BONOV, 1967). At the same time we are also taking into consideration the supersecular process of the solar activity.

References

ANDERSON, N.: J. Geophys. Res., T. 59 (1954).
BONOV, A. D.: Ann. of the Univ. of Sofia, Fac. of Physics, T. 59.
BONOV, A. D.: Soln. Dannie, No. 3 (1957).
BONOV, A. D.: Bull. VAGO, No. 25 (1959).
BONOV, A. D.: Soln. Dannie, No. 10 (1967).
CHRISTIAKOV, V. F.: Bull. VAGO, No. 25 (1959).
GNEVYSHEV, M. N., OL, A. I.: Astr. Phys. J. [in Russian] T. 25 (1948).
HALE, G. E., NICHOLSON, S.: Astrophys. J., T. **62** (1925).
OPENHEIM, S.: Astron. Nachr., 232 (1925).
WALDMEIER, M.: Astron. Mitt. Zürich, No. 133 (1935).

The Intensity of the Penumbra of Sunspots

By G. Eriksen and P. Maltby
Institute of Theoretical Astrophysics, University of Oslo, Norway

With 2 Figures

Abstract

The results of the observations of penumbral intensities obtained by pinhole photometers during 1967–1972 are presented. The pinhole photometers and the method of observation are briefly described.

Discussion of our data and comparison with the results given by other authors suggest that the reference published given by Wöhl *et al.* (1970) give too low intensity values, particularly for wavelengths below 0.550 μm.

I. Introduction

During several years the intensity of sunspots has been studied at the Oslo Solar Observatory using broad band pinhole photometers, or alternatively using the grating spectrograph. In this paper we shall consider the results of the measurements of penumbral intensities by pinhole photometers.

The tower telescope at the Oslo Solar Observatory was described by Brahde (1967). The off-axis mirror system with 0.30 m aperture and 29.55 m focal length is used for these measurements. The solar image ($\approx$27 cm in diameter) is focused on the aperture plate of the pinhole photometer.

Several versions of pinhole photometers have been used starting with a simple version described by Eriksen (1963), followed by a 3-channel photometer (Rödberg, 1966) and a 4-channel photometer (Maltby and Mykland, 1969). The pinhole photometers have been constructed by one of us (G. E.). Today the photometers listed in Table 1 are used.

II. The Pinhole Photometers and Methods of Observations

Let us consider the six-channel pinhole photometer. Light from the solar image focused on the aperture plate and passing through the pinhole (0.15 mm diameter) is dispersed in a double-pass prism spectrograph as shown in Fig. 1. Six different detectors well screened from stray light are placed at selected positions in the spectrum. The light beam is chopped at a rate of 600 Hz by means of a 12-bladed, motordriven disk. A 600 Hz reference voltage for the synchronous rectifier is derived from a silicon photovoltaic cell illuminated by a small lamp through the same chopper.

Table 1. Characteristics of the pinhole photometers

Type of photometer	Pinhole size (″)	Channel No.	Wave-length (μm)	Band-width (μm)	Detector
Six-channel	1.0 (may be increased to 10″ for measurements of aureole intensity)	1	1.670	0.080	Germanium, Siemens APY 11
		2	1.215	0.040	Germanium, Siemens APY 11
		3	0.876	0.020	Silicon, Siemens BPY 44
		4	0.669	0.010	Silicon, Siemens BPY 44
		5	0.579	0.006	Photomult. EMI 9609B
		6	0.387	0.002	Photomult. EMI 9609B
Double-channel	1.0	1	0.383	≈0.010	Photomult. 1P21
		2	0.330	≈0.024	Photomult. 1P21
Single channel	2.0	1	2.200	0.200	PbS-cell
Single channel	3.0	1	3.800	0.400	PbSe-cell
Single-channel	1.0	1	Centered on any wavelength between 0.350 and 0.600	0.004–0.025	Photomult. EMI 9609B

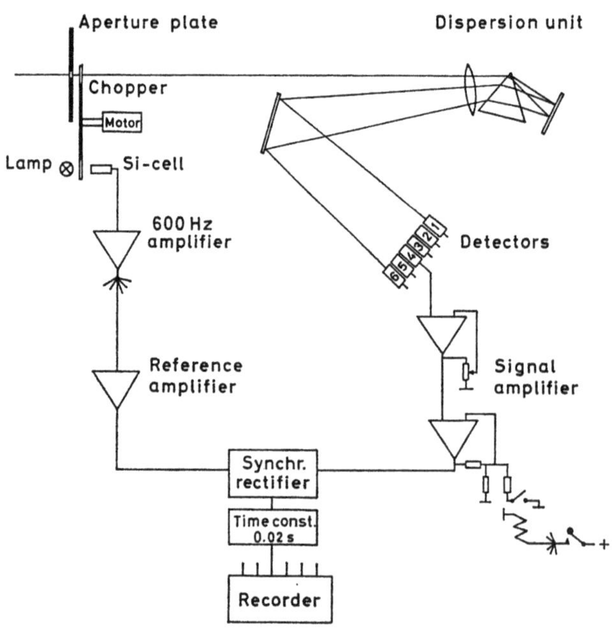

Fig. 1. Light path in the six-channel pinhole photometer and block diagram for one of the channels

In Fig. 1 is also shown a simplified diagram of the electronics of one of the six channels. The 600 Hz detector signal is amplified in two stages. The first stage has manual gain control for individual setting of the signal level appropriate for each channel. The second stage has means for increasing the gain of all channels by a fixed factor by relays operated by a button. The relays may also be operated automatically (not shown in Fig. 1) when recorder deflection decreases below a chosen level. This is convenient when measuring the aureole intensity. The bandwidths are determined by the width of a slit in front of each detector.

Similar detection techniques are used in the other pinhole photometers.

The sunspots are observed by letting the solar image drift across the pinhole aperture. During each observing series (unless interrupted by poor seeing conditions) the intensity in each point of the sunspot region will be measured by parallel scans through the image of the sunspot region. The intensities of the umbra and the penumbra are usually determined from scans going through the center of the spot. Accordingly, several scans through the central part of the spot are recorded. In addition several scans giving the intensity distribution at the solar limb as well as the intensity distribution of the aureole are observed.

The penumbral intensity is determined from scans where a penumbral plateau is apparent. The extent of the penumbral plateau is usually much smaller than the distance between the rims of the umbra and the penumbra. We avoid measuring in parts of the penumbra influenced by image blurring by selecting a fraction of the observed scans for further reduction.

III. The Penumbral Intensities

Some of the results of our measurements of penumbral intensities have already been published. MALTBY and MYKLAND (1969) gave the results for 11 spots observed in 1967 and 1968 for 2 wavelength regions in the visible as well as for 2 wave-length regions in the infrared. They selected 267–271 scans from each of the 4 wave-length regions. MALTBY and STAVELAND (1971) studied 1 sunspot in 8 wavelength regions between 0.360 μm and 4.0 μm. Recently, we have just completed the reduction of the data for the years 1968–1971 and the results are presented in Table 2. The penumbral intensities given in Table 2 are corrected for scattered light using a correction method suggested by MALTBY and STAVELAND (1972). The possible difference in line absorption in the penumbra and the photosphere is neglected.

In Table 2 is given the r.m.s. difference between the different observing series. It is found that the penumbra intensity may change from one spot to another in the sense that a low intensity in one wave length region implies a low intensity also at the other wave-lengths. We note that the penumbral intensity also alters with position within the spot; the r.m.s. intensity difference between opposite sides of the same spot is 2.9% of the photospheric intensity at 0.579 μm. Thus, the variation in penumbral intensity within the same spot is comparable with the change in intensity from one spot to another. In the comparison of different spots we have, for each spot used the average value derived from all selected scans. A more detailed account of the 1968–1971 observations is given elsewhere (MALTBY, 1972).

Table 2. Penumbral intensities $\langle I^*/I \rangle$ (equal weight to each observing series with $\geqq 4$ scans)

Wave-length (μm)	No. of spots	No. of obs. series	No. of scans selected	$\langle I^*/I \rangle$-r.m.s. deviation
0.387	19	39	338	0.640 ± 0.020
0.402 0.406	12	38	430	0.634 ± 0.043
0.450 0.455	7	17	155	0.714 ± 0.023
0.513	3	6	56	0.726 ± 0.031
0.579	49	121	1227	0.768 ± 0.022
0.669	49	122	1256	0.794 ± 0.018
0.876	49	121	1267	0.827 ± 0.017
1.215	18	37	323	0.876 ± 0.014
1.670	49	123	1287	0.914 ± 0.011
2.200	14	41	422	0.930 ± 0.008
3.800	4	16	207	0.936 ± 0.008

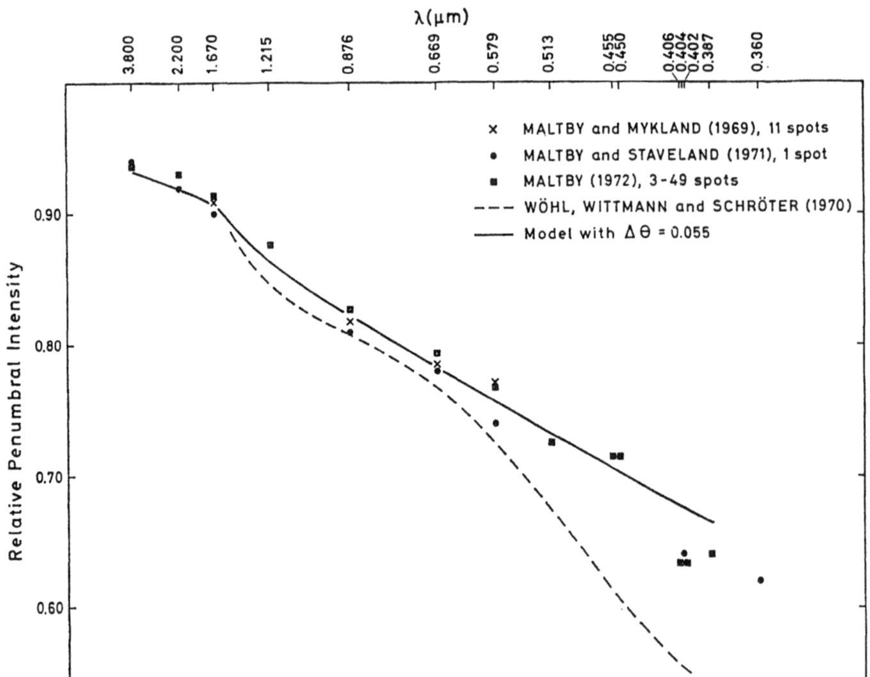

Fig. 2. The intensity of the penumbra relative to the photosphere versus the inverse of the wavelength. The notations used are shown in the diagram

WÖHL, WITTMAN, and SCHRÖTER (1970) observed the penumbral spectrum of 1 sunspot in the wave-length region 0.390–0.800 μm. Combining their results with values of penumbral intensities published by different observers prior to 1970 WÖHL *et al.* (1970) presented standard reference data for the penumbral intensity[1]. In Fig. 2 the data given by WÖHL *et al.* (1970) are shown by a dashed line. The results of the penumbral observations at the Oslo Solar Observatory are also given in Fig. 2.

It appears from Fig. 2 that the values given by WÖHL *et al.* (1970) are systematically lower than our values. Particularly for wave-lengths below 0.550 μm the deviation is large, up to 9% of the photospheric intensity at 0.400 μm. We note that the magnitude of the correction for scattered light we have applied to our data is much less than the difference between the two sets of data. If the line absorption is different in the penumbra and the photosphere, it is likely that the line absorption is largest in the penumbra. Thus, if we did correct our data for line absorption an even larger discrepancy would exist. The discrepancy cannot accordingly be explained by our correction for scattered light or by the lack of correction for line absorption. A closer examination of the data given by WÖHL *et al.* (1970) shows that their low intensity values for the penumbra between 0.400 and 0.550 μm is mainly based on their own observations of 1 sunspot, whereas higher values are given by MICHARD (1953) and by MAKITA and MORIMOTO (1960) from observations of several spots. We are therefore tempted to search for sources of error in the measurements of WÖHL *et al.* (1970). As their entrance slit is as large as 8.4 × 1.2 sec of arc, light from other parts of the spot than the small part of the penumbra that should be measured may enter the spectrograph.

In Fig. 2 are also shown the intensities corresponding to the $\Delta\theta = 0.055$ model of KJELDSETH MOE and MALTBY (1969). Preliminary results from observations by EKMANN and MALTBY indicate that the difference between this model and the observations may be significant in the violet region. Our observations in the violet region of the penumbral spectrum are made with a 2-channel pinhole photometer; the wave-length regions are centered on 0.330 and 0.383 μm. The presence of penumbral filaments also requires a refinement of the penumbra model.

References

BRAHDE, R.: Institute Theoretical Astrophysics, Report No. 23 (1967).
ERIKSEN, G.: Institute Theoretical Astrophysics, Report No. 15 (1963).
KJELDSETH MOE, O., MALTBY, P.: Solar Phys. **8**, 275 (1969).
MAKITA, M., MORIMOTO, M.: Publ. Astron. Soc. Japan **12**, 63 (1969).
MALTBY, P.: Solar Phys. (1972).
MALTBY, P., MYKLAND, N.: Solar Phys. **8**, 23 (1969).
MALTBY, P., STAVELAND, L.: Solar Phys. **18**, 443 (1971).
MALTBY, P., STAVELAND, L.: Institute Theoretical Astrophysics, Report No. 33 (1972).
MICHARD, R.: Ann. d'Astrophys. **16**, 217 (1953).
RÖDBERG, H.: Nature **211**, 394 (1966).
WÖHL, H., WITTMANN, A., SCHRÖTER, E. H.: Solar Phys. **13**, 104 (1970).

1 Dr. E. H. SCHRÖTER has kindly informed us that some of the numbers given in Table 5 of WÖHL *et al.* (1970) are incorrect. The correct values for the relative penumbral intensities are 0.798, 0.805, 0.811 and 0.817 for the wavelengths 0.800, 0.850, 0.900 and 0.950 μm, respectively.

Discussion

De Jager, C.:
How do you plan to account for the inhomogeneities in the model of penumbrae of sunspots?

Maltby, P.:
We hope to construct a two-component model in such a way that the model agrees with the observed intensities for all aspect angles.

Solar Flares, Suprathermal Plasma Nodules, and Loop Prominences

By L. D. DE FEITER and C. DE JAGER
The Astronomical Institute, Utrecht, The Netherlands

With 1 Figure

Abstract

We describe the small bright points visible on X-ray heliographs with fairly high resolution; we call them suprathermal plasma nodules. They seem to be due to the storage, in a magnetically confined region, of plasma ($\simeq 1$ keV) produced by a series of solar flares. If the storage energy exceeds a certain limit, the magnetic structure disrupts and expands, leading to features like loop prominences. This model can be confirmed by some numerical estimates.

I. Suprathermal Plasma Nodules

The concept of suprathermal plasma nodules was introduced by the present authors (DE FEITER and DE JAGER, 1973) from an investigation of a solar soft X-ray image described by BURGER and DIJKSTRA (1972). This picture, taken in nearly monochromatic light, shows half a dozen bright points with diameters not exceeding the spatial resolution of the camera (20″), clearly distinguishable from the more diffuse other X-ray structures visible on the heliograph. A striking fact is that their X-ray brightness increases with the number of flares that *has* occurred in the vicinity of the nodule. This leads to the suggestion that part of the hot plasma produced by a series of consecutive flares in an active region is accumulated and confined in

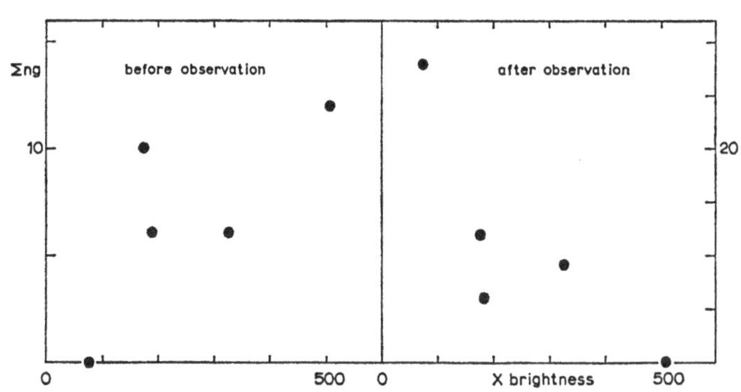

Fig. 1. Left: weighted numbers of flares that occurred in the plasma nodules before the X-ray observation, plotted against the X-ray brightness of the nodules. Right: the same for the flares that occurred in the nodules after the observation

the nodules. This process apparently continues until the energy content of the nodule exceeds a certain limit. There are indications that at that time disruptive processes take place leading to a change of the magnetic configuration. There are clear indications that after this disruptive process the probability for flares to occur in the active region reduces: actually the number of flares that appears to occur in the vicinity of the nodules is a decreasing function of the brightness hence of the apparent life time of the nodule. The brightest nodule did not produce any flare at all after the day of the observation.

In this paper we wish to attract attention to the disruptive process accompanying the instability of the nodule.

II. Relation to Loop Prominences and Flares

It is a striking fact that the brightest nodule, to which the greatest number of flares had contributed and in which no further flare occurred after the day of observation was related to a loop prominence. This is a strong empirical argument for examining the relation between the disruption of the magnetic structure of a nodule and loop-like structures. On many occasions BRUZEK (1962, 1964, 1969; BRUZEK and DEMASTUS, 1970) have drawn the attention to the relation between loop structures related to flares. The flare loop seems to develop as slow knots of material that grow and eventually form the loop structure. Downward streaming of material is observed at both ends of the loop. The velocities are of the order of tens to some hundred km/s. The loop expands with velocities of the order of ~ 10 km/s. The feet of the loops are not anchored in solar flares but rather in the active region, on either side of the so-called neutral line. There are various observations showing that these expanding arches leading to loop-like structures tend to emerge from the flare in its initial phase.

It is well known that the flare loops are closely related to sporadic coronal condensations, a concept introduced by WALDMEIER (1957). These are diffuse and highly excited coronal regions, occurring at the same place and having shapes similar to loop prominences.

There is also a close relation between the above described picture of expanding flare-associated loops and the model of a solar flare itself. In a model described by one of the authors (DE JAGER, 1969) a flare is pictured as a fairly flat loop (not rising above about 10000 km) anchored in the solar active region, on either side of the neutral line. The magnetic field of the flare itself may be twisted. In this model, supported by various observational data (e.g. MORETON and SEVERNY, 1968) the flare ignition is thought to be due to a current instability as suggested by ALFVÉN and CARLQVIST (1967) so that energetic material is produced in the loop system. It is now easy, and tempting, to link the above picture about solar flares with the observations of suprathermal plasma nodules and the flare associated coronal expanding loops: the more flares are produced in the flare's loop system, the greater the energy content of this closed system. Instability arises when the internal pressure of the suprathermal gas exceeds the tension of the curved magnetic system; then the loop starts expanding and convects upwards giving rise to the coronal loop system. Hot plasma material expands also into the corona and eventually into interplanetary space.

III. Some Energy Considerations

On the basis of the energies involved it can be shown that the above developed model is internally consistent. Table 1 gives some physical characteristics of nodules and sporadic coronal condensations (loop prominences). Data about the nodules has been taken from the paper by BURGER and DIJKSTRA (1972) and from that of REIDY *et al.* (1968). The X-ray emission of eruptive loop prominences has been studied by BRINKMAN and SHAW (1972) and we have also incorporated the eruptive source studied by BURGER and DIJKSTRA (1972). The volume V of the sources was derived from the emission measure as given by the authors, devided by N_e^2. The total energy content E of the source is computed with NkTV. It is striking that there is no substantial difference between the densities and volumes of the nodules and those of the loop prominences which may signify that the highly excited part of the loop prominence (responsible for the X-ray emission) is just the nodule lifted up by convective motions after the disruption of the magnetic structure. In our opinion this is an important observation which may be of significance for the further development of theories on the relation between flares and loop prominences and which also may shed light on the mechanism of the escape of energetic particles from the sun. It is also interesting to note that the disrupting nodule (column 6 of the Table 1) has an energy well above that of normal loop prominences and is a is multiple of the energy contained in the nodules. This again supports the idea that the disrupting nodule originates because of the accumulation of energy after the occurrence of several solar flares.

It is of interest to estimate the critical magnetic field of the nodule at which disruption takes place. In source A of BURGER and DIJKSTRA (column 6 of the Table 1) the internal pressure NkT$\approx$40 erg. cm^{-2}. The tension of the magnetic field lines necessary to compensate this pressure should be of the order of $B^2/4\pi$. This yields a limiting magnetic field of the order of 20 Gauss which does not look unreasonable.

Table 1. Physical characteristics of nodules and sporadic coronal condensations (loop prominences)

	Nodules			Loop prominences		
	1	2	3	4	5	6
T (10^6 K)	3*	3*	1.8	4	14	7*
Ne (10^{10} cm^{-3})	1.4	11	>0.7	2	10	3.7
V (10^{28} cm^3)	7	0.3	8	2	0.03	6
E (10^{29} erg)	4	1	2	2	0.5	18

* assumed. 1 source E of BURGER and DIJKSTRA (1972). 2 source J of ibid. 3 REIDY *et al.* (1968). 4 BRINKMAN and SHAW (1972), core. 5 Ibid, main structure. 6 source A of BURGER and DIJKSTRA (1972).

References

ALFVÉN, H., CARLQVIST, P.: Solar Phys. **1**, 220 (1967).
BRINKMAN, A. C., SHAW, M. L.: Solar Phys. **23**, 120 (1972).
BRUZEK, A.: Zeitschr. f. Astrophys. **54**, 225 (1962).
BRUZEK, A.: Astrophys. J. **140**, 746 (1964).
BRUZEK, A.: In: Solar flares and space research (C. DE JAGER and Z. SVESTKA, eds.), p. 61, Amsterdam, 1969.
BRUZEK, A., DEMASTUS, H. L.: Solar Phys. **12**, 447 (1970).
BURGER, M., DIJKSTRA, J. H.: Solar Phys. **24**, 395 (1972).
DE FEITER, L. D., DE JAGER, C.: Solar Phys., **28**, 183 (1973).
DE JAGER, C.: In: Solar flares and space research (C. DE JAGER and Z. SVESTKA. eds.), p. 1, Amsterdam 1969.
MORETON, G. E., SEVERNY, A. B.: Solar Phys. **3**, 282 (1968).
REIDY, W. P., VAIANA, G. S., ZEHNPFENNIG, T., GIACCONI, R.: Astrophys. J. **151**, 333 (1968).
WALDMEIER, M.: Die Sonnenkorona II, Basel 1957.

Interpretation of a Hard Solar X-Ray Burst

By P. HOYNG and G. A. STEVENS
The Astronomical Institute, Space Research Laboratory, Utrecht, The Netherlands

With 4 Figures

Abstract

We present an analysis of the flare of 18 May, 1972, UT 14:06, observed with the Utrecht hard solar X-ray spectrometer on board the ESRO satellite TD–1A. In this analysis we include the pertinent GHz radio data of this event. The proton density in the source region is inferred to be at least $5 \cdot 10^9 \text{ cm}^{-3}$. The radio spectra are found to agree with the calculations of HOLT and RAMATY. The radio and X-ray data together result in a consistent set of parameters.

I. Discussion of the X-Ray Data

On May 18, 1972 a 1B flare of X-ray class M4 occurred in McMath region 11883. The event started at UT 14:06, reached maximum intensity in $H\alpha$ at UT 14:09, and its area was $3.7 \cdot 10^{18} \text{ cm}^2$ [Solar Geophysical Data (prompt reports), June 1972]. No type III burst has been observed in connection with this flare. In fact, this event appears to be a complex event with a faint precursor flare, followed by a major event 5 degrees *SE* of the precursor (MCINTOSH, 1972). Fig. 1 displays time profiles of the X-ray burst associated with this event as observed with our hard solar X-ray detector (VAN BEEK and DE FEITER, 1973). In channel 5 (100–140 keV) and the higher channels the solar radiation did not exceed the instrumental background. The X-ray flux during the first 35 sec is probably due to the precursor event. The spectrum fits very well to a (steep) power law at all times, as illustrated in Fig. 2.

The interpretation is well known by now: electrons are accelerated in the solar atmosphere and injected into a region with a magnetic field and a certain ambient density, giving rise to synchrotron radiation (GHz radio burst) and Bremsstrahlung by electron-proton collisions (X-rays). The fast electrons loose their energy by electron-electron collisions. The average energy loss is given by (BROWN, 1971):

$$\frac{dE}{dt} = -\frac{55.7\,\pi e^4}{E}\, n_p v(:)\, n_p E^{-\frac{1}{2}} \tag{1}$$

where e, E and v are the electronic charge, kinetic energy and velocity; n_p is the ambient proton density.

For an initial electron spectrum of the form $N(E, 0)\, dE = c E^{-\gamma}\, dE$ we define a half-life time $t_{\frac{1}{2}}$ for decay due to collisions [as given by formula (1)] by

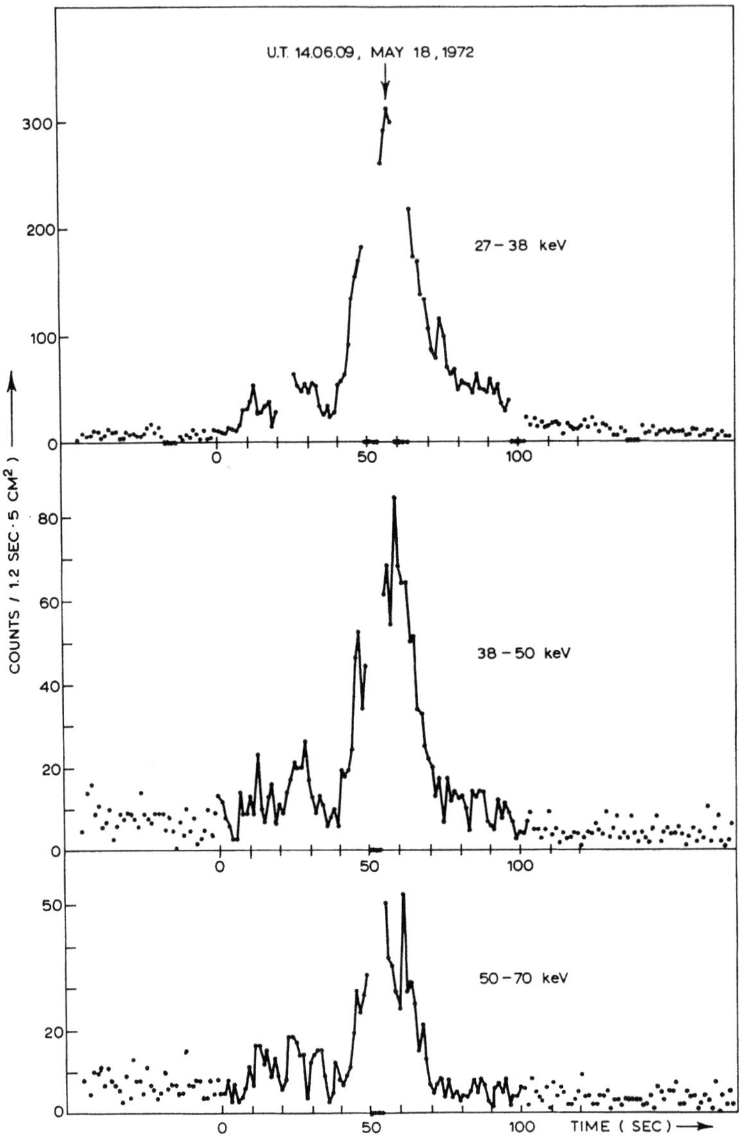

Fig. 1. Contents of channels 1, 2 and 3 as a function of time. Missing data are due to bad telemetry and for channel 1 also to subcommutation

$N(E, t+t_{\frac{1}{2}})=\frac{1}{2}\, N(E, t)$. The following expression is derived:

$$t_{\frac{1}{2}}=\left(t+\frac{10^8\, E_{\mathrm{keV}}^{3/2}}{n_p}\right)\cdot\left(2^{3/(2\gamma+1)}-1\right). \qquad (2)$$

The time variations in Fig. 1 are too big to be of statistical origin (e.g. up to $\sim 4\sigma$ in channel 3). A lower limit to the ambient density is found from formula (2),

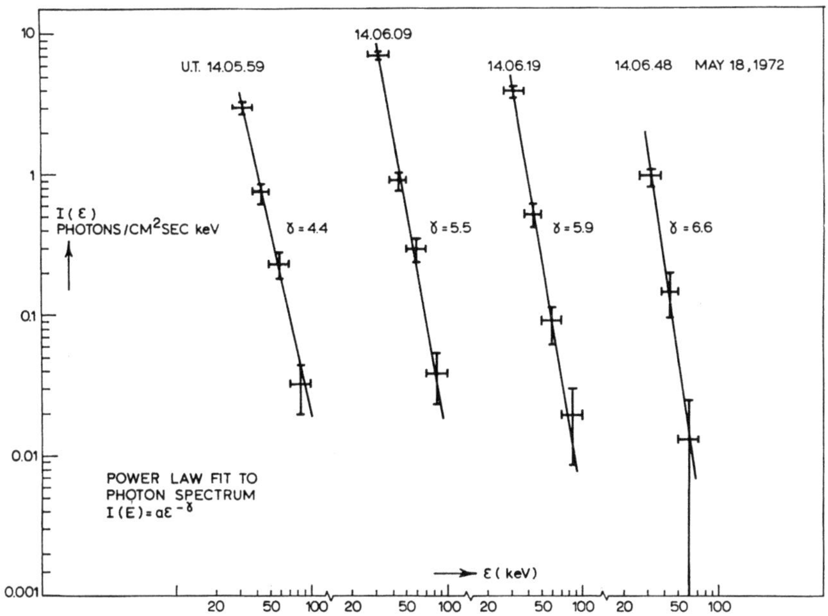

Fig. 2. Power law fit to the X-ray spectrum at four different times. Background is subtracted

noticing that the decay time at ~ 60 keV is less than 2 sec:

$$n_p \geqq 5 \cdot 10^9 \,\text{cm}^{-3}.$$

The actual density in the source region is probably much higher; as the time resolution of hard solar X-ray detectors improved during the last ten years, they continued to detect time variability in the X-ray flux up to the limit set by their time resolution. In all our observations up to now we found these fast-decay fine structures. Although more easily tractable, we feel however that the so-called impulsive injection models are untenable.

We can compute the electron injection function $f(E)$, if we take the photon spectrum $I(\varepsilon)$ to be a power law at any instant ($I(\varepsilon) = a\,\varepsilon_{\text{keV}}^{-\gamma}$ phot. $(\text{cm}^2 \cdot \text{sec} \cdot \text{keV})^{-1}$) and if we assume the ambient density to be sufficiently above its lower limit (thick target assumption; Brown, 1971):

$$f(E) = 7.34 \cdot 10^{33} \cdot a\gamma(\gamma-1)^2 \frac{\Gamma(\gamma-\frac{1}{2})}{\Gamma(\gamma)} E_{\text{keV}}^{-(\gamma+1)} \,\text{electr}\,(\text{keV} \cdot \text{sec})^{-1} \tag{3}$$

$\Gamma(x)$ is the gamma function.

Thus we find the total injected energy due to electrons with $E \geqq 27$ keV between $t=0$ and $t=100$ sec to be $1.5 \cdot 10^{30}$ erg; the total number of these injected electrons: $3 \cdot 10^{37}$. The maximum injection rates occur at UT 14:06:09 viz. $6.3 \cdot 10^{28}$ erg/sec and $1.2 \cdot 10^{36}$ electrons/sec, respectively.

In section 2 we need the total number of electrons in the source region, N, with energies above 100 keV, at $t = t_0 + 15$ sec. By extrapolation of the injection function, this number is estimated to be $N \simeq 10^{34}$. In this calculation decay effects are accounted for, taking $n_p = 5 \cdot 10^9$ cm^{-3}.

II. Radio Data and their Relation to the X-Ray Data

In Fig. 3 the microwave emission at several frequencies associated with this flare is shown. The shape of this burst exhibits the same features as seen in Fig. 1. The only conspicuous difference is the long-lasting tail in the radio flux density. At

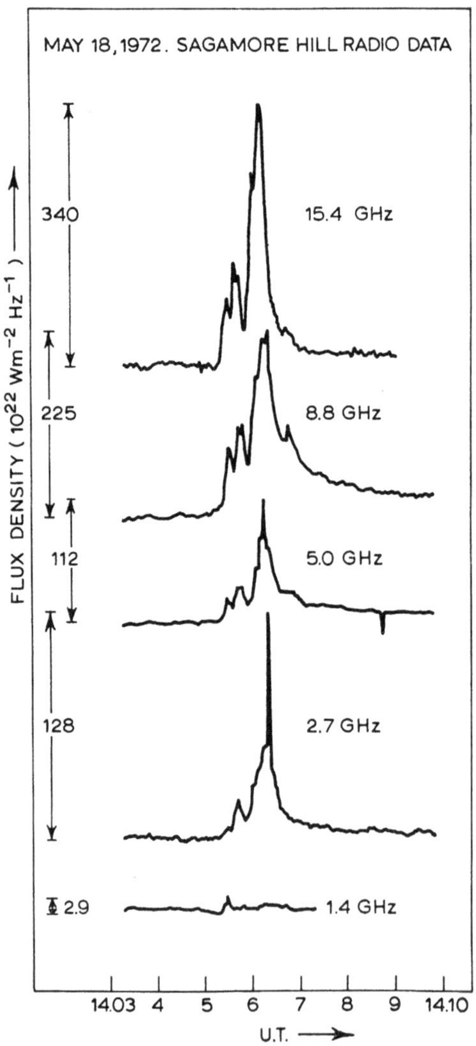

Fig. 3. Time profiles of the microwave flux density at five different frequencies associated with the flare of 18 May, 1972, UT 14:06

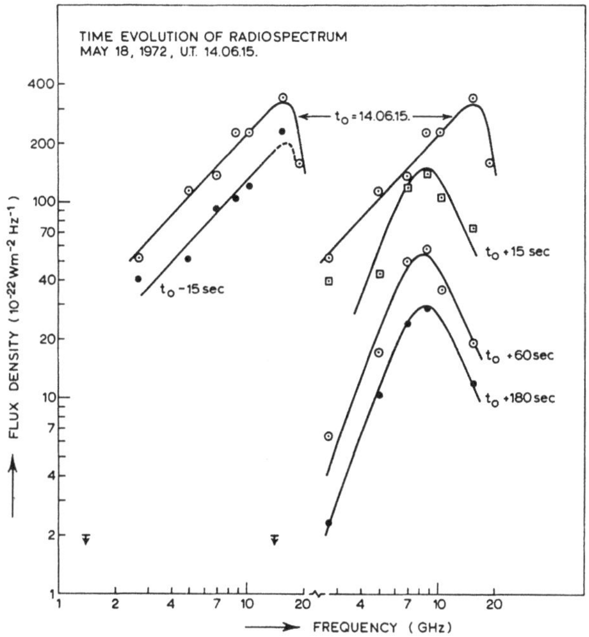

Fig. 4. Microwave spectra at five different times. The curves at $t=t_0+15$, t_0+60 and t_0+180 sec are spectra calculated by Holt and Ramaty (1969) and the other two are hand drawn best fits

8.8 GHz the decay of this tail obeys:

$$F(8.8\,\text{GHz})(:)(t-t_0)^{-2/3} \tag{4}$$

giving a decay time of ~ 15 sec at $t=t_0+15$ sec. This may be compared to values derived from formula (2) as follows: Assume a reasonable magnetic field, say 100 G, then 8.8 GHz corresponds to emission in the ~ 20th harmonic. As indicated by Holt and Ramaty (1969) one then needs electrons of relatively high energy, say $E \gtrsim 100$ keV. According to formula (2), taking $n_p = 5 \cdot 10^9\ \text{cm}^{-3}$ and $t=0$, one needs energies of the order ~ 150 keV.

So we think that the tail is due to electrons with energies $\gtrsim 100$ keV slowly decaying in the source region.

There remains the problem that these electrons do not give detectable X-ray flux, so we must derive their total number.

Fig. 4 shows the time evolution of the radio spectrum.

The spectra at $t=t_0+15$, t_0+60 and t_0+180 sec show good agreement with the Holt and Ramaty calculations. Surprisingly, this fit is better than for the flare studied by Holt and Ramaty (1969) themselves and it contradicts the remarks made by Zirin *et al.* (1971) on the steepness of the Holt and Ramaty spectra.

Unfortunately, the *H & R* spectra assume a low energy cut-off at 100 keV in the electron spectrum. So a comparison of the spectra for $t \leqq t_0$ is impossible due to the copious presence of electrons with energies below 100 keV.

A quantitative, though not exclusive, fit of the spectrum at $t=t_0+15$ sec with $n_p=5\cdot10^9\ \mathrm{cm}^{-3}$ yields the following values for the magnetic field B, the area A of the emitting region and the total number N of electrons with energies above 100 keV:

$$B=95\ \mathrm{G};\qquad A=1.3\cdot10^{17}\,\mathrm{cm}^2;\qquad N=10^{34}.$$

The spectra do not depend strongly on H & R's parameter α, so we took their value $\alpha\simeq0.7$. The value of N is the same as derived above. Since 10^{34} electrons $E\geqq100$ keV, given the steepness of their spectrum and the background levels, cannot give any detectable X-ray flux, we can say that the X-ray and microwave data confirm each other. A weak point is that the value of γ used by H & R does not apply. It is therefore desirable that their calculations are extended to other values of γ and lower energy cut-offs. The area A is by a factor 20 smaller than the maximum $H\alpha$ area; they apply, however, to times separated by ~3 min.

Acknowledgements

The radio data included in this analysis have been kindly placed at our disposal by Drs. CASTELLI (Air Force Cambridge Research Laboratories), CROOM (Radio and Space Research Station, Slough), KAUFMANN (Centro de Radio-Astronomia e Astrofisica, Sao Paulo), and SCHANDA (Institut für Angewandte Physik, Bern).

References

BEEK, H. F. VAN, DE FEITER, L. D.: This volume, page 103, 1973.
BROWN, J. C.: Solar Phys. **18**, 489 (1971).
HOLT, S. S., RAMATY, R.: Solar Phys. **8**, 119 (1969).
MCINTOSH, P. S.: 1972, private communication.
ZIRIN, H., PRUSS, G., VORPAHL, J.: Solar Phys. **19**, 463 (1971).

The Hard Solar X-Ray Spectrometer on Board the ESRO Td-1a Satellite

By H. F. van Beek and L. D. de Feiter
The Astronomical Institute, Space Research Laboratory Utrecht, The Netherlands

With 5 Figures

Abstract

The solar hard X-ray spectrometer on board the ESRO TD-1 A satellite (launched on March 12, 1972) aims at measuring X-rays in the energy range from 24 to 900 keV. This energy range is divided into twelve approximately logarithmically spaced energy channels.

Continuous pointing at the sun provides the possibility for a fine time resolution, which has been set at 1.2 sec for the 4 channels between 24 and 90 keV and 4.8 sec for the others. Careful rejection of background radiation and very accurate preflight and inflight calibration add to the reliability of the observations. The paper describes the instrument and its characteristics.

Introduction

The satellite was launched into a near-polar, sun-synchronous, circular orbit at an altitude of 550 km. Due to the precession of the orbital plane, sunshine on the satellite along its whole orbit is maintained for at least seven months. After a hibernation period of about three months the satellite may start a new observing sequence, early February 1973. Apart from six other experiments among which one for the detection of solar gamma rays, the TD satellite contains the spectrometer described hereafter. It is especially designed to analyse, with good time resolution, the spectral distribution of photons in the energy range from 24 to 900 keV emitted by certain types of solar flares. The viewing direction of the instrument is along the axis of the satellite which is kept continuously pointed towards the Sun.

The instrument was switched on and worked properly since March 18, 1972.

General Instrument Description

A general view is given in Fig. 1 and a schematic outline of it is presented in Fig. 2.

The instrument consists essentially of a scintillation counter with a sensitive area of 5 cm^2. The CsI (Na) crystal has a thickness of 15 mm and is optically coupled to a EMR photomultiplier tube. Impulse height analysis is done by

Fig. 1. The instrument S 100 during preflight calibration

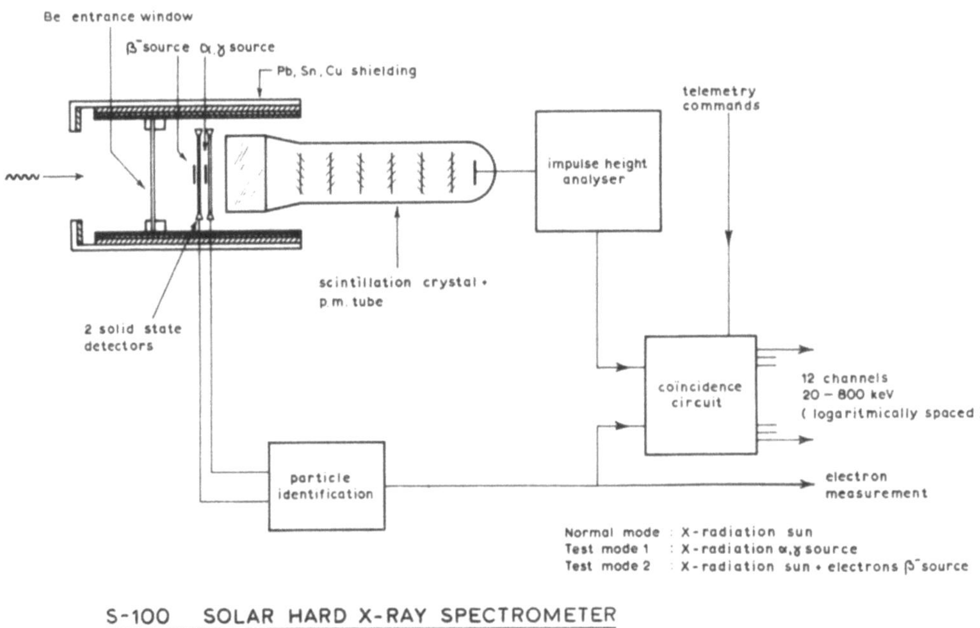

Fig. 2

means of discriminators because of low dead time and redundancy advantages in comparison with a multichannel analysing system. The energy range is split up into twelve logarithmically spaced channels, which is just sufficient for the analysis of the photon spectral distribution. The channel limits are 24, 34, 45, 63, 90, 125, 175, 270, 320, 420, 540, 670 and 900 keV. The sampling rate is 1.2 sec for the four low-energy channels and 4.8 sec for the others.

Special care has been taken to minimize the effects of background radiation on the count rates in the scintillation counter. This background radiation consists mainly of X- and gamma rays, resulting from interaction of cosmic-ray particles with the satellite material, and cosmic X-rays. During the passes of the satellite through the radiation belts and through the South Atlantic Anomaly the background is enhanced due to electrons. These electrons cause counts both by penetration into the scintillation crystal as well as by the generation of Bremsstrahlung in the satellite material close to the detector. To absorb the background X-rays of various sources a lead-tin-copper shielding around the scintillation counter, ending in a collimator in front of the counter, has been applied. The full width at half maximum of the collimator is 26°. Inside the collimator and in front of the scintillator an entrance window of beryllium (thickness 1.2 mm), and two solid-state-detectors (thickness 350 microns each) have been mounted. All particles entering the scintillator within the field of view will also be detected by the two solid-state detectors; this way pulses in the scintillation counter due to these particles can be rejected.

The total thickness of the silicon solid-state detectors could be chosen in such a way that a suitable efficiency cut-off at lower energies (15 keV) results in order to prevent the scintillation counter from saturation effects during strong solar flares. The use of two solid-state detectors instead of one enables us to apply a coincidence measurement between these detectors, performed in the particle identifier.

The purpose of this is to prevent the solid-state detector noise impulses from contributing to the scintillation counter dead time. For in-flight calibration two radioactive sources, ^{90}Sr and ^{241}Am, have been mounted in front of and between the two solid-state detectors respectively. On ground command coincidence measurements between the solid-state detectors and the scintillation counter can be initiated. In these calibration modes of operation, only the photon spectrum of the ^{241}Am source or the electron spectrum of the ^{90}Sr source will be analysed, thus providing an accurate measurement during the flight, of the efficiency, the resolution and amplification of the scintillation counter, and of the discriminator levels. These measurements are performed a few times per week.

In the normal mode electrons from the Sr source are treated just as cosmic particles and Van Allen belt electrons and therefore will be rejected.

In coincidence with photons the Am source emits alpha particles which will penetrate into one of the solid-state detectors and thus will activate the anti-coincidence circuit. By this means the photons emitted by this radioactive source are almost completely rejected in the normal mode of peration.

Some technical figures: weight 4.4 kg, power consumption 3.3 Watt, bit rate $\sim$50 bits/sec.

Preflight Calibrations

For the determination of the overall efficiency and the spectral resolution of the instrument, as well as for the investigation of secondary effects, measurements of several X-ray and gamma ray lines from radioactive sources have been made with the instrument prior to its launch. In total fifteen lines, ranging from 14.36 keV (^{57}Co) to 834.8 keV (^{54}Mn), were selected for this purpose. From these measurements the detector characteristics have been determined over the entire energy range with an accuracy within a few percent. For the final data reduction, two instrumental effects are of prime importance: the escape peak, which affects the measurements in the range from about 33 keV to 120 keV, and the Compton effect which shows up if the primary photon energy exceeds about 180 keV.

Theoretical considerations suggest that because of the use of a CsI (Na) crystal the escape peak in fact consists of two peaks; one caused by the escape of a CsK-photon out of the crystal, the other initiated in case an I K-photon leaves the crystal.

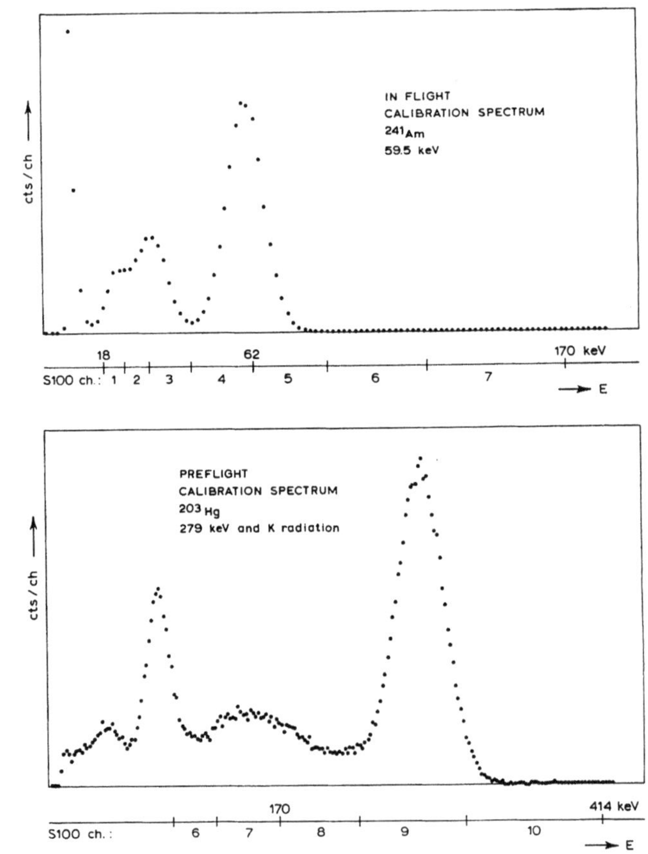

Fig. 3. The detector response for two monoenergetic photon energies

The effects due to Compton scattering can be split up into two components, one results from Compton scattering in the crystal giving rise to the generally known Compton distribution. Compton scattering of photons just outside the crystal introduces a second distribution of pulses, with quite a different spectral distribution. The integrated intensity as a function of the primary photon energy, however, has the same character. A last small effect is caused by a very thin dead layer of Cs I which acts as an additional absorber in front of the detector. About half of the photons absorbed in this layer lead to Cs and I K-radiation entering the active part of the scintillation crystal. Fig. 3 gives two examples of the detector response. In Fig. 4 the relevant efficiencies of the instrument as a function of the energy are presented. Because most solar flares exhibit a fairly steep spectrum, and all effects mentioned are known in detail, the original photon distribution can be accurately retrieved from the observations.

The spectral resolution of the instrument as a function of energy was also derived from the measurements of mono-energetic X- and gamma lines. It can be represented by the formula:

$$R(E) = 6.0 + \frac{125}{\sqrt{E}}\ \%;$$

R is the spectral resolution (FWHM) and E the energy in keV. This means that the resolution is 26% at 40 keV and 11% at 660 keV.

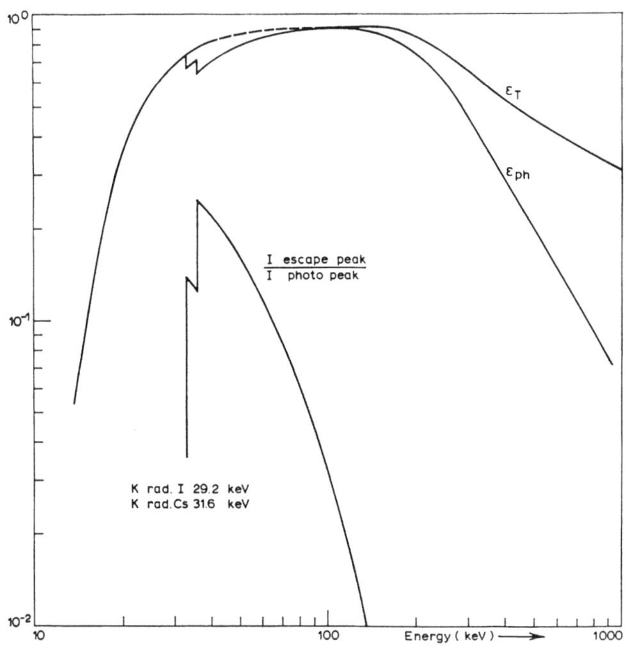

Fig. 4. Instrument efficiencies

Observations of the Instrument in Orbit

As an example of the counting rates which occur in orbit, two time profiles have been given here, showing the contents of channel 1 as a function of time.

Fig. 5a displays the count rate for a representative orbit. A little flare (April 18, 1972, 01:01 U.T.) has been observed while the instrument was just above the

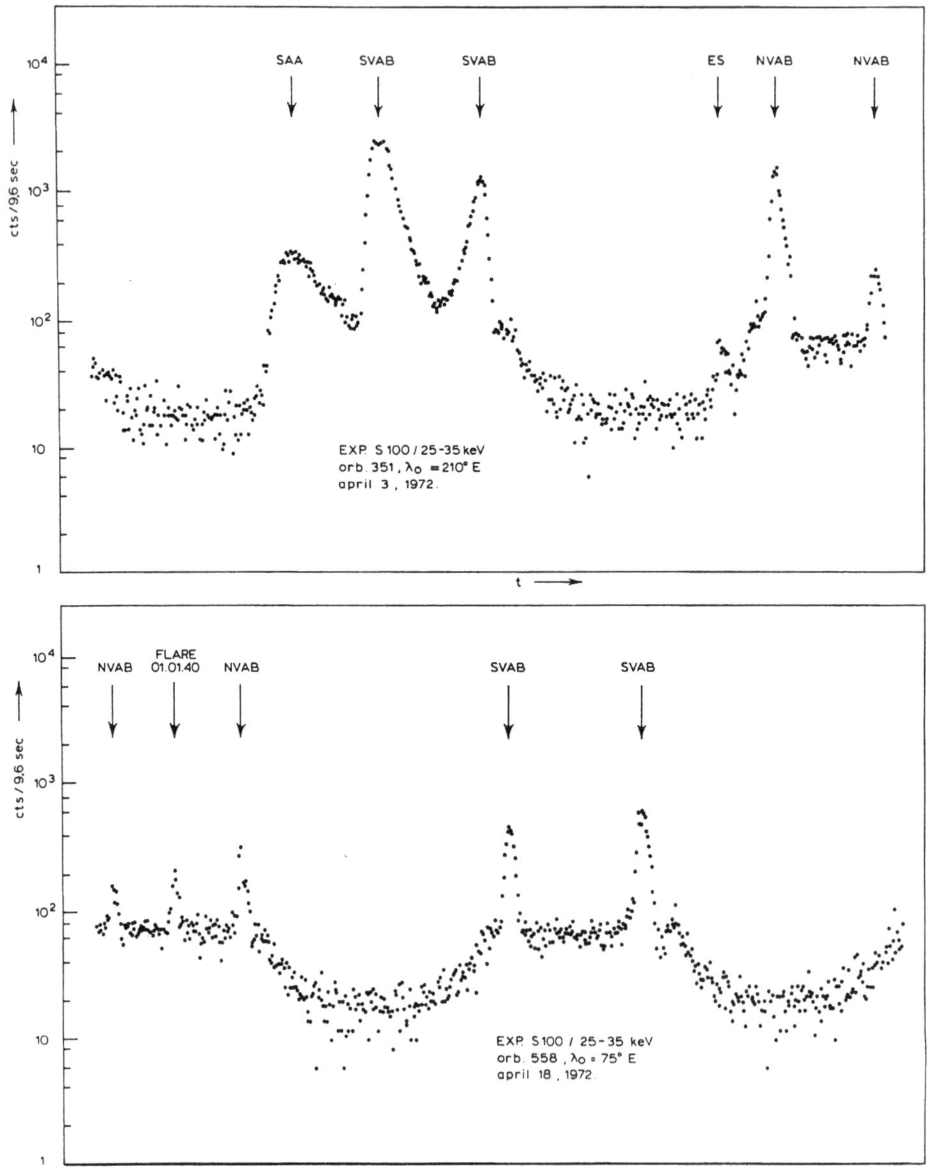

Fig. 5a and b. Count rates for channel 1 during two typical orbits

north pole. During the orbit shown in Fig. 5b the experiment crosses the South Atlantic Anomaly; during that time, just as during Van Allen belt passes, the instrument is of course less sensitive and can only observe strong solar flares.

At 43° N and 173° E a curious feature was observed, which also appeared on previous and following orbits crossing about the same point. It is indicated in the figure as electron spot (E.S.). Further analysis of this and similar structures in the background radiation awaits the availability of a complete set of data.

Although the instrument was designed to cover also the high energy range (several hundreds of keV's) the optimalization for lowest background still turns out to be successful as appears from the low background count rates outside the radiation belt regions.

Detailled Photometrical Study of the Development of a Flare from Hα Observations

By R. FALCIANI and M. RIGUTTI
Capodimonte Observatory, Naples, Italy

Several solar flares have been photographed in the light of the Hα line by C. J. MACRIS at the National Observatory in Athens. Some two thousands Hα filtergrams have been analyzed getting—for each filtergram—from six to ten isophotes at fixed intensity levels and with these development curves were obtained.

The discussion of these curves led to some new information about the behaviour of flares of importance from 1 to 2. The most important information are:

1) the nature of an Hα flare is that of a pulsating phenomenon;

2) the initial phase of the phenomenon is often a rapid contraction;

3) the maximum of emission is rapidly reached (flash-phase) after the contraction;

4) the emission grows then fainter exponentially with time, going through a series of pulsations of increasing duration; the rate of the exponential decrease is an important parameter. When it is low, successive pulsations may bring the emission to values larger than that at the moment of the flash. This fact may influence the definition of the moment of the maximum phase. Sometimes a fluctuation superimposed on a slow decrease flare may induce to believe that a new flare develops at the same spot;

5) the analytical representations for both the pulsation durations and the exponential decrease are the same for flares showing strong contraction and flash-phase, as well as for flares showing weak contraction and flash-phase;

6) all the aspects of a flare analyzed in this paper are more and more pronounced as the considered intensity level is higher and higher.

Other general features of the flare development are the following:

1) during the contraction phase the whole region except the one where the flash will develop decreases in intensity (and in area);

2) during the flash-phase, small regions appear and rapidly develop at high intensity levels;

3) after the flash-phase a diffusion of the most intense regions towards lower intensity levels take place and the photometrical structure becomes simpler;

4) the pulsations after the flash-phase take place in the same region; this fact excludes a mechanism of propagation of the disturbance from one region to another;

5) the final photometrical configuration of the active region is like the initial configuration; this shows that a flare such as the those we analyzed does not destroy the general structure of the active region.

At present the possible correlations between the Hα features development and the X and radio emissions at different wave-lengths are considered. One first result seems to be that the presence of higher and higher intensity $H\alpha$ levels in the optical flare is correlated to the presence of X-ray emissions at shorter and shorter wavelengths. The optical area of a flare might then not be very important in order that the flare should be an X-ray source.

Discussion

DE FEITER, L. D.:
Did you compare your interesting results with the EUV observation made by NOYES *et al.* on board OSO VII? These latter observations refer to the layers in between the $H\alpha$ flare and the X-ray region.

FALCIANI, R.:
No, because our present $H\alpha$ data refer to a period before OSO VI observations.

Peculiar Finestructures in Solar Radio Spectra of Type IV Emission

By A. D. Fokker
Utrecht University Observatory 'Sonnenborgh', Utrecht, The Netherlands

With 5 Figures

Abstract

Peculiar patterns of fine structural details in the continuum emission of type IV solar radio-bursts have been observed at great contrast with the 60-channel radiospectrograph of the Utrecht Observatory. A short description of these phenomena and some comments are given.

Type IV radiation is a broad band continuum radio emission associated with major flares. It has components in the centimetric, the decimetric and the metric frequency ranges which have different characteristics.

In the decimetric type IV continuum finestructures have first been reported by Young *et al.* (1961). These authors recognized an intermediate drift type of emission with a bandwidth of a few MHz that drifts through the spectrum towards lower frequencies. Finestructures also occur in the higher metric frequencies. Elgarøy (1961) reported a type of modulation pattern that consists of parallel drifting bands at around 200 MHz.

Due to the high intensities of the type IV continuum, conventional radiospectrographs are often saturated, with the result that relatively weak fine-structural details appear inconspicuous. In the 60-channel solar radio spectrograph of Utrecht Observatory the background, averaged over the immediate past 3 sec, is subtracted. The instrument thus displays the short-lived intensity fluctuations relative to a floating reference level by which the contrast is much enhanced. In this way, finestructural modulations of type IV emission have been observed in considerable detail on a number of occasions. Five types have been noted so far:

a) "Fiber bursts" (identical to the intermediate drift bursts, see Fig. 1). Very strikingly, a fiber burst generally consists of an emission ridge with an adjacent absorption edge on the low frequency side of the emission. The fiber bursts thus suggest that radiation is transferred over an interval of 2 or 3 MHz from one frequency to a slightly higher frequency.

b) Pulsating structure, consisting of relatively broad band (several tens of MHz) quasi-periodic intensity fluctuations. The quasi-period typically amounts to 0.5 to 2 sec. The fluctuations are generally almost in phase over their full bandwidth; sometimes they show a slight tilt in the frequency-time plane (see Fig. 2).

c) Broad band absorptions. Dark strips of $\ll 1$ sec to 1 or 2 sec duration, giving the impression that part of the continuum emission, over a bandwidth of several tens of MHz, is suddenly switched off and on (see Fig. 3).

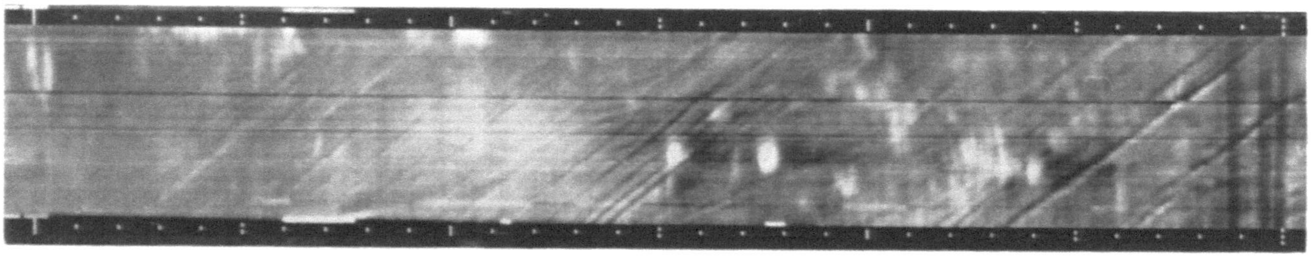

Fig. 1. Intermediate ("fiber") bursts on June 29, 1971, recorded by the Utrecht radiospectrograph. The frequency range is from 253 MHz (bottom) to 213 MHz (top). Time runs from left to right; seconds are marked

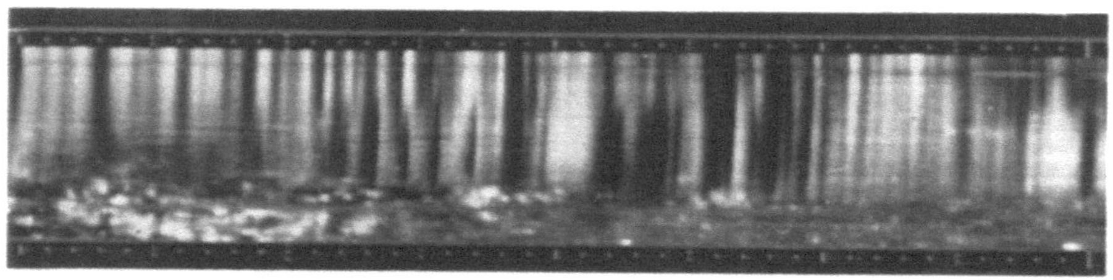

Fig. 2. Pulsating structure, recorded by the Utrecht radiospectrograph on February 25, 1969. The frequency range is from 320 MHz (bottom) to 160 MHz (top). Time runs from left to right; seconds are marked

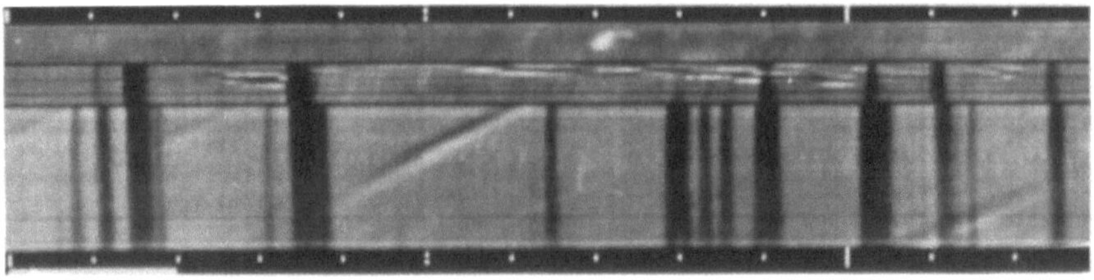

Fig. 3. Broad band absorptions on March 6, 1972, recorded by the Utrecht radiospectrograph. The record covers three frequency bands: 315–282 MHz (lower), 246–234 MHz (middle) and 199–187 MHz (upper). Time runs from left to right; seconds are marked

d) "Zebra patterns" (identical to the parallel drifting bursts), consisting of several approximately equidistant ($\Delta f = 2$ à 5 MHz) narrow band ridges (see Fig. 4). A given pattern persists typically during 2 to 6 sec. Within one pattern the drift rate commonly is variable.

e) "Tadpole structures". These are the most intriguing finestructures, so far only observed on one occasion, during an event on March 2, 1970. They consist of a deep absorption body, a bright narrow band emission dot ("eye") at the high frequency side and a relatively weak emission tail at the low frequency side. The structure is oriented nearly along the frequency axis, the bandwidth of body and tail is typically 40 MHz, the duration 0.2 to 0.3 sec. Maybe the "eye" is similar to the "small knots of radiation" with $\Delta f \lesssim 0.3$ MHz and $\Delta t \leqq 0.1$ sec as observed by ELGARØY (1961) in an event on 18 Aug. 1959. A detailed phenomenology of tadpole structures has been given by SLOTTJE (1972) (see Fig. 5).

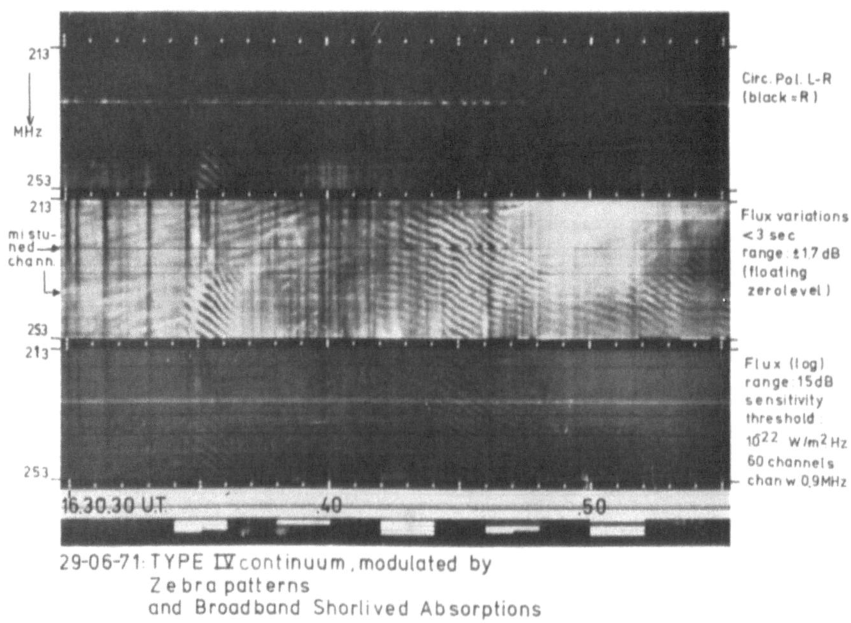

Fig. 4.

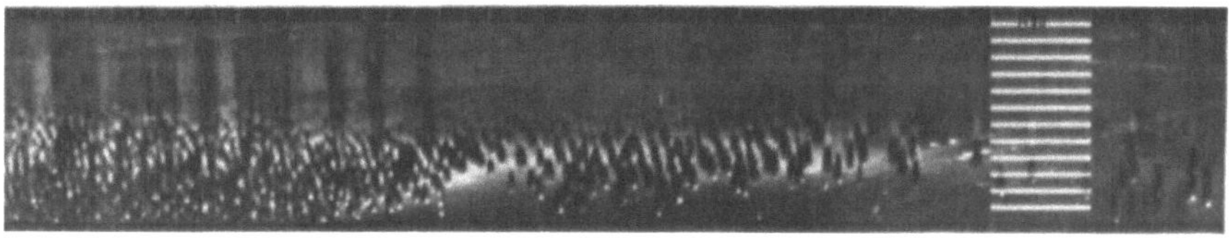

Fig. 5. Tadpole structures, recorded by the Utrecht radiospectrograph on March 2, 1970. The frequency range is from 320 MHz (bottom) to 160 MHz (top). Time runs from left to right; seconds are marked

On several occasions the real solar origin of these features was confirmed on comparison with fast running single frequency records taken at Trieste (327 MHz, see ABRAMI, 1972) and at Kiel (240 MHz).

The origin of these finestructures is largely still a mystery. The pulsating structure phenomenon was interpreted by ROSENBERG (1970) as evidence for the occurrence of magneto-hydrodynamic pulsations of magnetic flux tubes. Synchroton radiation emitted by fast electrons would be modulated by periodic variations in the magnetic field strength. MCLEAN *et al.* (1971) suggested that the fast electrons become concentrated at the top of a magnetic arch and that radial oscillations are due to the passage of a shock front.

The frequency separation Δf in zebra patterns is identified by ROSENBERG (1972) with the gyrofrequency. The idea is that radiation is emitted at the sum frequencies of the upper hybrid frequency and at several harmonics of the gyrofrequency.

The fiber bursts suggest that up-conversion of a wave must takes place by wave coupling and the interaction with a low frequency wave phenomenon, which might be a whistler, as suggested by J. KUIJPERS (private information). The phenomenology of fiber bursts has been treated by KUIJPERS and SLOTTJE.

The broad band absorptions need not be absorptions at all; they may just correspond with an interruption of part of the emission process. From the fact that radiation disappears over a substantial range of frequencies, we may infer that this emission is generated in a volume of very limited extent. Otherwise it is hard to understand how the blocking out could be so extremely sudden and practically simultaneous. Likewise, the regularity of spacing between the ridges in zebra patterns seems to require a well-defined parameter value (possibly magnetic field strength) which again would mean that the whole phenomenon is generated in a relatively small volume. An intriguing problem then is, how the required very high densities of emission can be generated. One may think of mechanisms of stimulated emission and on the possibility of strong directional beaming.

The desire to know more about the spatial properties of the finestructure of type IV emission calls for high resolution interferometric observations.

References

ABRAMI, A.: Nature, Phys. Sc. **238**, 25 (July 10, 1972).
ELGARØY, Ø.: Astrophys. Norv. **7**, 123 (1961).
KUIJPERS, J., SLOTTJE, C.: In preparation.

McLEAN, D. J., SHERIDAN, K. V., STEWART, R. T., WILD, J. P.: Nature **234**, 140 (1971).
ROSENBERG, H.: Astron. & Astrophys. **9**, 159 (1970).
ROSENBERG, H.: Solar Phys. **25**, 188 (1972).
SLOTTJE, C.: Solar Phys. **25**, 210 (1972).
YOUNG, C. W., SPENCER, C. L., MORETON, G. E., ROBERTS, J. A.: Astrop. J. **133**, 243 (1961).

Discussion

KARABIN, M.:
1. What is the percentage of occurence of that peculiar radio emission together with all type IV bursts?
2. If whistler mode of e.m. propagation is an explanation for that peculiar radio emissions how can we answer the question of coupling the whistler frequencies and escaping frequencies (peculiar emission).
Could you find any connection with a certain kind of a flare?

FOKKER, A. D.:
1. Fiber (or intermediate drift) bursts are not uncommon among type IV events. I estimate that at least 10% of the type IV produce them. If one includes the cases of very small modulation, the percentage may be considerelly higher.
2. The frequency of the whistler mode is supposed to be much smaller than the frequency of the longitudinal plasma waves with which they couple. The whistlers themselves cannot escape from the corona. Mr. KUIJPER'S idea is that the fiber burst is an indirect evidence of their existence.

RIGUTTI, M.:
May I ask you how much limited might be the volume of the solar atmosphere responsible for the peculiar absorption – emission features you observed and how much energy may be involved in this phenomenon?

FOKKER, A. D.:
A crude estimate of the size involved perhaps can he made by multiplying the presumed velocity of the whistler pulse (5000 km/sec, say) with the typical duration of the fiber, 0.1 to 0.2 sec. One then arrives at a size of hardly more than 1000 km. The diameter of the oscillating flux tube held responsible for pulsating structure was estimated by ROSENBERG at something like 1000 km. as well. The energy involved depends on the efficiency of the coupling conversion process. This point is being investigated by Mr. KUIJPERS.

Unusual Burst of Activity Associated with McMath Region 10789 on June 13, 14, 1970[1]

By A. D. FOKKER
Utrecht University Observatory 'Sonnenborgh', The Netherlands

With 2 Figures

Abstract

The solar radio burst activity associated with McMath Region 10789 on June 13, 14, 1970 was a subject of cooperative study initiated by the Committee of European Solar Radio Astronomers (CARRIGAN *et al.*, 1973). The findings are here shortly summarized.

The center of activity McMath region 10789 was remarkable for a most concentrated series of flare-associated radio events which happened on June 13 and 14, 1970. This *CA* was connected with an *E*-type sunspot at 18° N (C.M.P. on June 17.2).

Most of the optical flares of this series were only sub flares. There occurred one truly big flare event, on June 14, 13.21 U.T. Most of the radio events were mo-

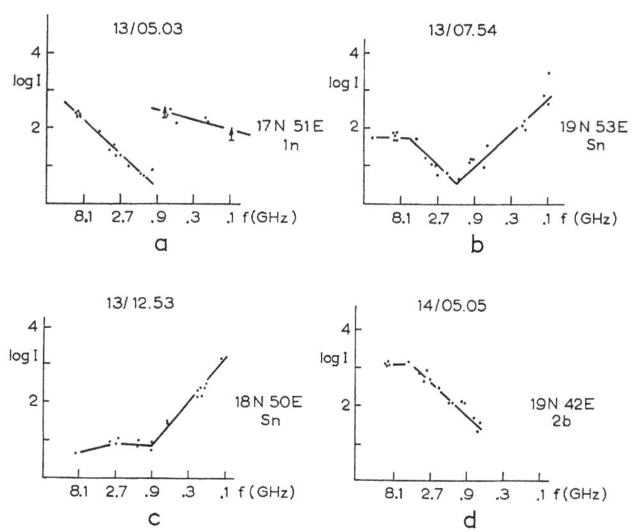

Fig. 1. Spectral plots of four radio events belonging to the flare series. *a* Jump-type connection; *b* *V*-type connection; *c* weak microwave burst, strong metric burst; *d* strong microwave burst, no metric response

1 This paper is part of a cooperative study of this period of activity, initiated by the Committee of European Solar Radio Astronomers (CESRA).

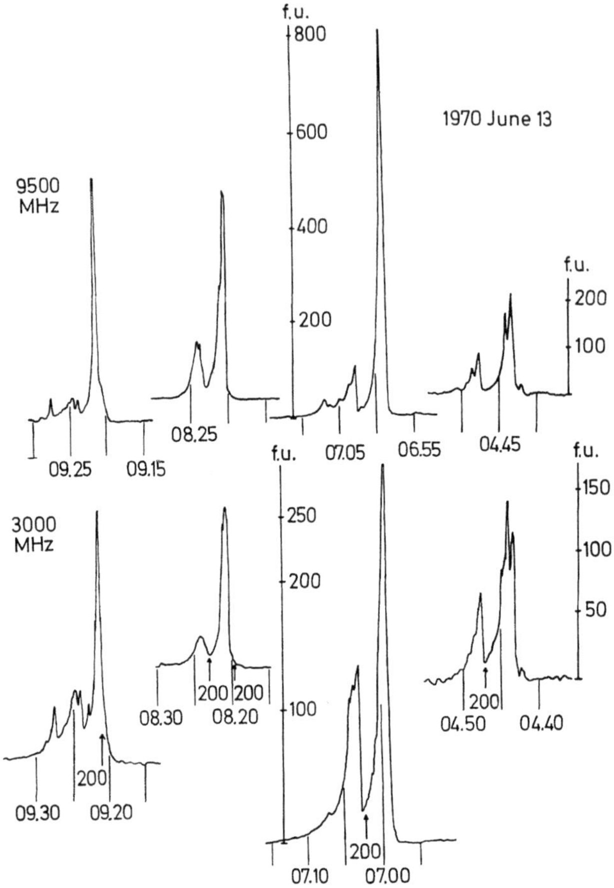

Fig. 2. Nera Observatory records at 3000 and 9000 MHz of four impulsive microwave events. The events of 06.58, 08.19 and 09.19 U.T. were homologous. The starting times of concomitant 200 MHz responses are indicated

derately sized. Very many sudden ionospheric disturbances were produced by flares in this CA (27 on June 13!).

This series of radio events is instructive, because it shows that conditions in a CA, during a limited time, may be favourable for the repeated and frequent occurrence of certain kinds of instability which again and again give rise to radio events. Among these events striking similarities are found as well as notable differences. Some of the features of activity in this CA are the following:

a) The radio productivity of the optical flares that occurred on June 13 is about $2\frac{1}{2}$ times the normal value and ranks among the highest rates encountered among flare families (cf. DE FEITER and FOKKER, 1961).

b) Various kinds of spectral plots (=log-log diagram of peak intensity versus frequency) are found amoung the radio events. Centimetric and metric branches connect in quite different ways: both V-type and *J*(ump)-type connections occur. Some events produce only a weak microwave burst but a strong metric response, other events (particularly later during the series) behave the other way (see Fig. 1). This is in line with what is true for radio events in general, namely that the microwave and metric responses of an event are largely independent of each other (cf. FOKKER, 1969).

c) In spite of the great variety of radio events, the number of sets of two events that were so much similar that they may be considered homologous, was as large as 7, ranking among the highest rates of homology found so far (cf. FOKKER, 1967). A certain number of events had the common property of producing very large responses around 100 MHz and lower frequencies. This property is typical for certain flare families. Another characteristic feature of certain flare families is the occurrence of impulsive microwave outbursts. The present CA was outstanding in the production of such bursts (see Fig. 2). The concentration of 9 impulsive microwave bursts in one and the same CA in an interval of less than $1\frac{1}{2}$ days is quite exceptional (cf. FOKKER and ROOSEN, 1961).

From this instance we learn the following:

a) A center of activity may go through a phase of instability of 1 or 2 days duration (for instance if magnetic flux emerges in the middle of an existing spot group, as pointed out by H. ZIRIN (personal communication).

b) Judging from the various kinds of radio events produced, discharges occurred at a few places where the magnetic configurations were obviously different.

c) The sun produces very sporadically certain exceptional series of flare events. This is a reason to keep constantly prepared for detailed observation of forthcoming occasions.

References

CARRIGAN, A. L., CASTELLI, J. P., ELGARØY, Ø., FOKKER, A. D., SCHEERER, H., URBARZ, H., LEBLANC, Y.: Proc. 3rd Meeting of CESRA, p. 94, Obs. de Bordeaux 1973.

DE FEITER, L. D., FOKKER, A. D.: *Bull. Astron. Inst. Neth.* **XV**, 319 (1961).

FOKKER, A. D.: *Solar Phys.* **2**, 316 (1967).

FOKKER, A. D.: *Solar Phys.* **8**, 376 (1969).

FOKKER, A. D., ROOSEN, J.: *Bull. Astron. Inst. Neth.* **XVI**, 83 (1961).

Ionization and Recombination Processes for Non-Thermal Plasmas

By M. LANDINI, B. C. MONSIGNORI FOSSI, and R. PALLAVICINI
Arcetri Astrophysical Observatory, Florence, Italy

Abstract

X-ray emission from solar flares in the spectral region 1–10 keV is usually interpreted as being produced by a thermal plasma at several million degrees of temperature. Anyhow isothermal models of flares cannot explain line emission at these photon energies and multi temperature models are required (MEEKINS *et al.*, 1970; NEUPERT, 1971). Furthermore from broad-band ion-chamber measurements, indications exist that also at low-energies some X-ray bursts may be non-thermal in nature (LANDINI *et al.*, 1972), or at least a non-thermal emission component may be present (KAHLER and KREPLIN, 1971). For these reasons ionization and recombination processes have been investigated for a plasma whose electrons follow a non-maxwellian energy distribution.

The energy spectrum which has been assumed for electrons is:

$$dN(E) = K N_e E^{-\gamma} dE \qquad \text{for } E > E_1$$

where E is the electron energy, $N_e(\text{cm}^{-3})$ is the non-thermal electron density, K and γ are constants. A low-energy cut-off E_1 has been assumed and K is determined by the normalization condition:

$$\int_{E_1}^{\infty} dN(E) = N_e.$$

The processes which have been taken into account are collisional ionization, both radiative and dielectronic recombination and autoionization.

To evaluate the collisional ionization rate from a $(z-1)$ to a z times ionized atom the cross-section given by NOCI *et al.* (1971) has been used and the following relations have been obtained for integer values of $(\gamma + 1/2)$:

$$q_{\text{coll}} = 2.75 \cdot 10^{-10} (\gamma - 1) \frac{\xi}{E_1^{1/2} I_{z-1}} \cdot \left\{ \frac{1}{\gamma - 1/2} - 1.197 F_{(\gamma+1/2)} \left(0.18 \frac{E_1}{I_{z-1}} \right) \right\} \text{cm}^3 \text{sec}^{-1} \qquad \text{for } E_1 > I_{z-1}$$

$$q_{\text{coll}} = 2.75 \cdot 10^{-10} (\gamma - 1) E_1^{\gamma-1} \frac{\xi}{I_{z-1}^{(\gamma+1/2)}} \cdot \left\{ \frac{1}{\gamma - 1/2} - 1.197 F_{(\gamma+1/2)} (0.18) \right\} \text{cm}^3 \text{sec}^{-1} \qquad \text{for } E_1 < I_{z-1}$$

where I_{z-1} is the ionization potential of the $X^{+(z-1)}$ ion, ξ is the number of electrons in the shell from which ionization takes place and $F_{(\gamma+1/2)}$ is the exponential integral function of $(\gamma+1/2)$ order.

The rate of radiative recombination from a z to a $(z-1)$ ionized atom has been evaluated by using the atomic cross-section given by SPITZER (1948):

$$\alpha_R=7.295\cdot 10^{-17}(\gamma-1)\,E_1^{(\gamma-1)}\,\bar{g}\,z^4(I_H z^2)^{-(\gamma+1/2)}$$
$$\cdot\sum_n\left[\frac{1}{n}\int_{\varepsilon_1}^{\infty}\frac{\varepsilon^{-(\gamma+1/2)}}{(1+n^2\varepsilon)}\,d\varepsilon\right]\ \mathrm{cm}^3\,\mathrm{sec}^{-1}$$

where

$$\varepsilon=\frac{E}{I_H z^2},\qquad \varepsilon_1=\frac{E_1}{I_H z^2},$$

I_H is the hydrogen ionization potential, $\bar{g}$ is the Gaunt factor for free-bound transitions and has been assumed equal to unity. The sum is made over the n levels at which recombination occurs.

For the dielectronic recombination rate a simple formula has been developed following the method suggested for the thermal case by TUCKER and GOULD (1966). For recombinations to the $X^{+(z-1)}$ ion one obtains:

$$\alpha_{di}(i,\mathrm{tot})=2.02\cdot 10^{-12}(\gamma-1)\,E_1^{(\gamma-1)}\,z^{6/5}$$
$$\cdot\sum_{i'}(\bar{g}_{ii'})^{3/5}f_{ii'}\,W_{ii'}^{-(\gamma+0.3)}\ \mathrm{cm}^3\,\mathrm{sec}^{-1}\qquad \text{for } W_{ii'}>E_1$$

where $\bar{g}_{ii'}, f_{ii'}, W_{ii'}$ are respectively the mean Gaunt factor, the absorption oscillator strength and the excitation energy for the transition from the i level to the i' level.

The autoionization rate from a $(z-1)$ to a z times ionized atom has been evaluated by using the collisional excitation cross-section given by VAN REGEMORTER (1960). The computed rate is:

$$q_{\mathrm{aut}}(i,i')=4.43\cdot 10^{-10}\frac{(\gamma-1)}{(\gamma-1/2)}E_1^{(\gamma-1)}\,\bar{g}_{ii'}f_{ii'}\,W_{ii'}^{-(\gamma+1/2)}\,\mathrm{cm}^3\,\mathrm{sec}^{-1}\qquad \text{for } E_1<W_{ii'}$$

$$q_{\mathrm{aut}}(i,i')=4.43\cdot 10^{-10}\frac{(\gamma-1)}{(\gamma-1/2)}E_1^{-1/2}\,\bar{g}_{ii'}f_{ii'}\,W_{ii'}^{-1}\,\mathrm{cm}^3\,\mathrm{sec}^{-1}\qquad \text{for } E_1>W_{ii'}$$

where $\bar{g}_{ii'}$, $f_{ii'}$ and $W_{ii'}$ are respectively the mean Gaunt factor, the oscillator strength and the excitation potential for the first level above the first ionization limit. This level as well as the value of the oscillator strength have been chosen following JORDAN (1969).

In all the above given expressions for the rates the energy in measured in keV.

Numerical computations have been performed for the isoelectronic sequences of Hydrogen, Helium and Lithium, and for ions from Neon to Nickel. Values of the spectral index γ from 1.5 to 5.5 and of the low-energy cut-off E_1 from 0.5 to 10 keV have been taken into account.

Using the evaluated rates, the relative populations of H-like to He-like and of He-like to Li-like ions have been computed assuming that also during solar flares a quasi-steady state exists.

The intensity ratios between the resonance lines pertaining to two successive ionization stages of the same element have been evaluated for the above mentioned ions, in the assumption that the upper level is populated by collisions inside the non-thermal plasma.

A comparison has been made with some experimental intensity ratios observed during solar flares and reported in the literature (MEEKINS *et al.*, 1968; NEUPERT, 1971; DOSCHEK *et al.*, 1971).

The few available measurements are consistent with a cut-off $E_1 = 1$ keV and with a narrow range of γ values, at about $\gamma \approx 2.9$, while a thermal interpretation of the same experimental data requires a much wider range of temperatures from about $20 \cdot 10^6$ to $40 \cdot 10^6$ K. However the data found in the literature are very few and further observations of line intensities are needed to test whether the non-thermal processes are really working during solar flares.

References

DOSCHEK, G. A., MEEKINS, J. F., KREPLIN, R. W., CHUBB, T. A., FRIEDMAN, H.: Astrophys. J. **170**, 573 (1971).

JORDAN, C.: Mont. Not. R. astr. Soc. **142**, 499 (1969).

KAHLER, S. W., KREPLIN, R. W.: Astrophys. *J.* **168**, 531 (1971).

LANDINI, M., MONSIGNORI FOSSI, B. C., PALLAVICINI, R.: Solar Phys. (in press) (1972).

MEEKINS, J. F., DOSCHEK, G. A., FRIEDMAN, H., CHUBB, T. A., KREPLIN, R. W.: Solar Phys. **13**, 198 (1970).

MEEKINS, J. F., KREPLIN, R. W., CHUBB, T. A., FRIEDMAN, H.: Science **162**, 891 (1968).

NEUPERT, W. M.: In: C. MACRIS (ed.), Physics of the solar corona, p. 237, Dordrecht-Holland: Reidel Publ. Co. 1971.

NOCI, G. G., CANTÙ, A. M., CHIUDERI DRAGO, F. G., GRIFONI MANTELLINI, L. A.: Contract AF 61 (052) 67 C 0042, Arcetri Astrophys. Observatory 1971.

REGEMORTER, H. VAN: Mont. Not. R. astr. Soc. **121**, 213 (1960).

SPITZER, L., JR.: Astrophys. J. **107**, 6 (1948).

TUCKER, W. H., GOULD, R. J.: Astrophys. J. **144**, 244 (1966).

Discussion

HOYNG, P.:
Did you suppose all the electrons to have a power law distribution or only a fraction of them, the others having e.g. a maxwellian distribution?

PALAVICINI, R.:
We are considering ions of high atomic number and of very high ionization stages. These ions produce lines which are strong only during solar flares. For this reason a thermal contribution will be small and involve only non-thermal electrons.

Some Details of the Supergranular Motion

By Aleksandar Kubičela
Astronomical Observatory, Belgrade, Yugoslavia

With 6 Figures

Abstract

The low-noise (±62 m/s) spectrographic evaluation of supergranular velocities reveals: a) The sight-line velocity distribution within the supergranules which is explained by the presence of Coriolis velocities, and b) Some places where, after an enhancement of approaching sight-line velocities, a typical supergranular velocity distribution evolves; this is probably how the cells are born. The Ca II K_{232} emission in most cases follows the supergranular flow.

Since the definition of the supergranulation has been established (Leighton *et al.*, 1962; Simon and Leighton, 1964), several observers have investigated the nature of the supergranular motion (e.g.: Tanenbaum *et al.*, 1969; Musman and Rust, 1970; Frazier, 1970; Deubner, 1971). The connection of the supergranulation with granular motion, magnetic fields, chromospheric and photospheric network, as well as the evolution of sunspots, has also been extensively studied by these and other investigators.

The statistical approach to supergranular velocities has been used by most of the early observers. However, as the data concerning velocities and evolutional changes of individual supergranules were missing, an extensive spectroscopic approach to the problem has been attempted in Kodaikanal during 1969.

The Observations

The observations were made with the large Kodaikanal solar spectrograph (11 mm/Å at the observed line Ba II 4554 Å). Using out-of-focus photospheric spectra as references, the supergranular velocities have been observed relatively, in respect to a very large scale mean sight-line velocity field in the same region of the photosphere. Three photospheric regions 139″ × 79″ were observed from three to six times. About 50 supergranules have been identified. The sample errors of the obtained velocities are between ±36 m/s and ±82 m/s (mean: ±62 m/s). Besides, Ca II K_{232} synthetic spectroheliograms of the same photospheric regions have been obtained by means of the same instrument about one hour after the velocity observations. The analyzed fields are listed in Table 1. Here the quantities θ_0, φ_0 and λ_0 represent heliocentric angle and heliographic coordinates of the centre of the corresponding fields.

Table 1. Some of the Observed Fields

Field No.	Date (1969)	U.T. h	min	Θ_0 (°)	φ_0 (°)	λ_0 (°)	Figure No.
A 1013	Jan. 4	3	24	50	47 S	28 E	2 (detail)
A 1014	Jan. 4	5	48	49	47 S	26 E	2 (detail) and 4
A 1015	Jan. 5	3	37	44	47 S	16 E	1, 2 (details) and 5
A 1028	Febr. 16	2	55	52	58 S	13 E	6
A 1062	March 31	3	06	46	52 S	01 E	3 (detail)
A 1063	March 31	6	17	46	52 S	00	3 (detail)
A 1064	March 31	9	16	46	52 S	02 W	3 (detail)

Tangential Component of the Supergranular Motion

Some systematic effects have been noticed. The most interesting is the position of the middle zero-velocity diameter of supergranular cells. Considering a simple horizontal flow diverging from a centre, one would expect that the bisecting sight-line zero-velocity diameter of a cell occupies the orthogonal position with respect to the radius of the solar disc (LEIGHTON *et al.*, 1962). However, the average angle between middle zero-velocity diameters of 32 observed cells and the radius of the solar disc amounts to 80°. The phenomenon can be explained by the presence of tangential (perpendicular to supergranule radius) component in supergranular horizontal motion. The direction, as well as the order of magnitude of the observed component correspond to the direction of Coriolis acceleration and Coriolis velocity found for couples of typical supergranules.

As an example, the observed sight-line velocities of supergranule No. 9 have been shown in Fig. 1a. Direction to the centre of the solar disc is C_1AC. Full circles represent the observed approaching sight-line velocities, and the open circles represent the receding velocities. Dotted curve MM_1 is the middle zero-velocity line (diameter). The mean intensities of the observed sight-line velocities

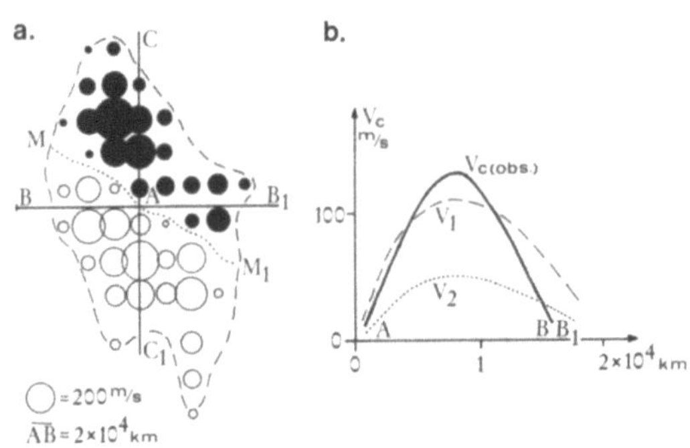

Fig. 1a and b. Coriolis velocities in supergranule No. 9. a) The observed sight-line velocities: full circles figure approaching motion, open circles figure receding motion. b) Velocity distribution along AB and AB_1: $V_{c(\text{obs.})}$ = the observed velocities; V_1 and V_2 = calculated Coriolis velocities

along AB and AB_1 are shown as heavy line, V_c(obs.), in Fig. 1b. The Coriolis sight-line velocities along the radii AB and AB_1 have been calculated, assuming the same supergranular horizontal velocity distribution along BB_1 as along CC_1. Further, it was assumed that the observed velocity field had already lasted for 20 hours (maximal useful time interval between the two observations) or 8 hours (the shortest reasonable life-time for such a developed supergranule). Using these assumptions, the Coriolis sight-line velocities have been calculated and shown as dashed, V_1 for $\Delta t = 20$ h, and dotted, V_2 for $\Delta t = 8$ h, curves in Fig. 1b. As it can be seen, the Coriolis velocity hypothesis under still reasonable assumptions represents the observed velocities quite well.

Similar results have been obtained from three observations of another supergranule (No. 12).

Some Evolutional Changes of the Velocity Field

In cases of some supergranules (No. 6 and No. 43) an abrupt local enhancement of approaching velocities and a subsequent typical supergranular velocity picture has been observed.

For example, supergranule No. 6 is a smaller-than-average cell with moderate but typical supergranular velocities in Fig. 2a. The approximate centre of the supergranule is marked with S. The representation of velocities and orientation of supergranules are the same as in Fig. 1.

The point of interest, indicated with X, is near the border of the cell, about 10000 km to the left from its centre (Fig. 2a). Only 2.4 hours after the first observation, approaching velocities appeared abruptly at X (Fig. 2b). Still about 22 hours later (Fig. 2c), a new typical supergranular sight-line velocity distribution can be seen around the position X (which is now a zero-velocity region).

Similarly, following a "normal" situation near the boundary of supergranule No. 33, at X in Fig. 3a, strong approaching velocities have been observed at the

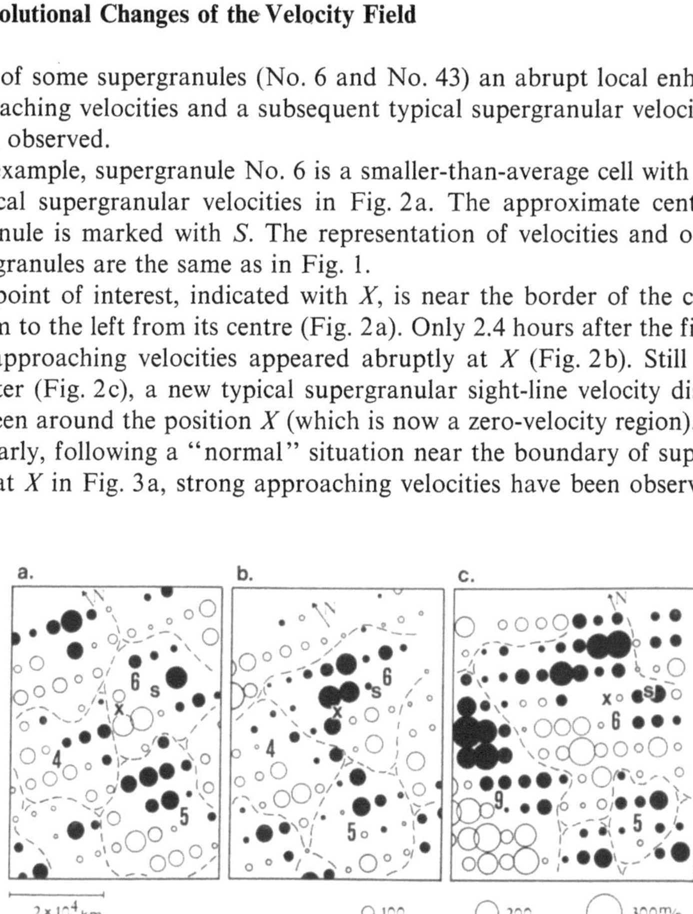

Fig. 2a—c. Velocity changes in supergranule No. 6. a) The observation A 1013, b) A 1014 and c) A 1015. Velocity represented as in Fig. 1. Dashed lines are supergranular boundaries. The position references are the points S and arrows N (North). The centre of the solar disc is upward

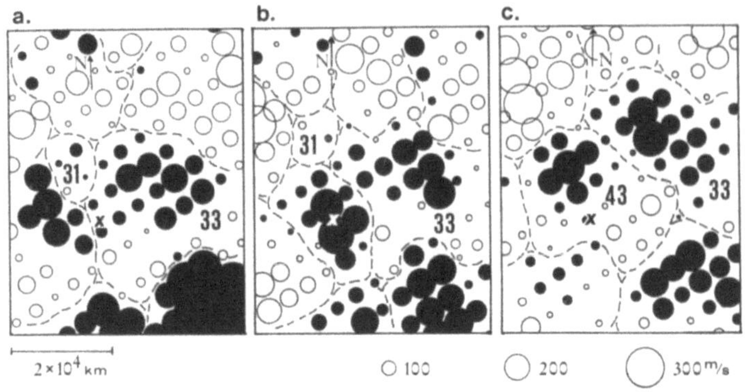

Fig. 3a–c. The formation of the supergranule No. 43. a) The observation A 1062, b) A 1063 and c) A 1064. Velocities shown as in Fig. 1. Dashed lines = supergranular boundaries. The position references are the points X and arrows N (North). The centre of the solar disc is upward

same point about 3.2 hours later (Fig. 3b). After another interval of 3.0 hours, a new supergranule No. 43 can be recognized around the position X (Fig. 3c).

If the observed enhanced sight-line velocities in Fig. 2b and Fig. 3b have to be interpreted as vertical upward velocities, they amount to 230 m/s with an acceleration of at least 95 m/s · h and 260 m/s with 80 m/s · h respectively.

Ca II K_{232} Emission Network

Ca II emission configurations mainly follow the supergranular motion. Some examples can be seen in Figs. 4b and 5b where the reconstructed Ca II K_{232} spectroheliograms are shown as isophotes in arbitrary intensity units. Comparing them with the corresponding velocity fields, Figs. 4a and 5a, one can notice the following changes: 1) The distribution of the isophotes C_1, C_2, C_3 and C_4 near (or along) the boundary of supergranule No. 1, 2) Decay of the isophote E at the centre of a new supergranule (No. 9) and its redistribution as E_1 and E_2, and 3) Changes in isophotes F_1, F_2 and F_4 after the velocity event in supergranule No. 6.

There are also some exceptions. Namely, the stable isophote A moved slightly between the observations A 1014 and A 1015. The reasons for such a movement, as well as the velocity changes at the boundary of supergranules No. 1 and No. 8, Fig. 5, are not very clear. However both events might be depending on the unknown configuration of the magnetic field at this position.

A more outstanding exception in Ca II emission behavior is the formation of a temporary emission "bridge" over supergranule No. 15 (Fig. 6). There was only a slight tendency of formation of the emission M a day before, and a slender trace of it a day after the observation A 1028. As the phenomenon shows apparently no connection with the velocity structure of supergranule No. 15, one might consider it as a high-level chromospheric feature (incomplete and unfinished "filling up" of the emission network interspace).

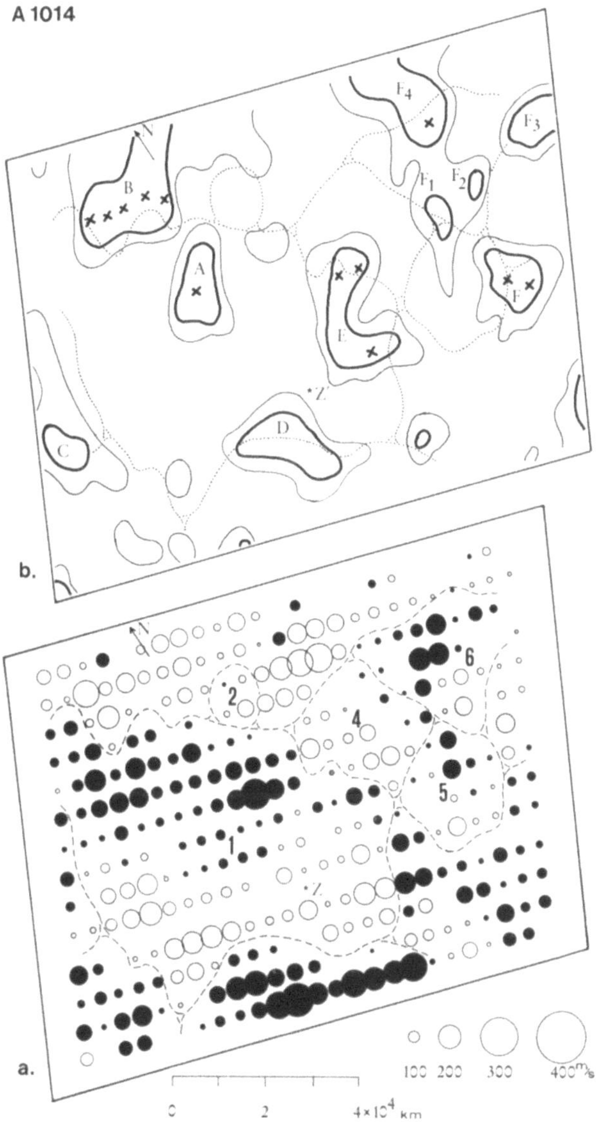

Fig. 4a and b. The complete observation of the field A 1014. a) Velocity field shown as in Fig. 1. Dashed lines figure supergranular boundaries. b) Ca II K_{232} intensity isophotes (arbitrary units), X are points of maximum intensity. Dotted lines figure supergranular boundaries repeated from a). The position references are points Z or Z' and arrows N (North). The centre of the solar disc is upward

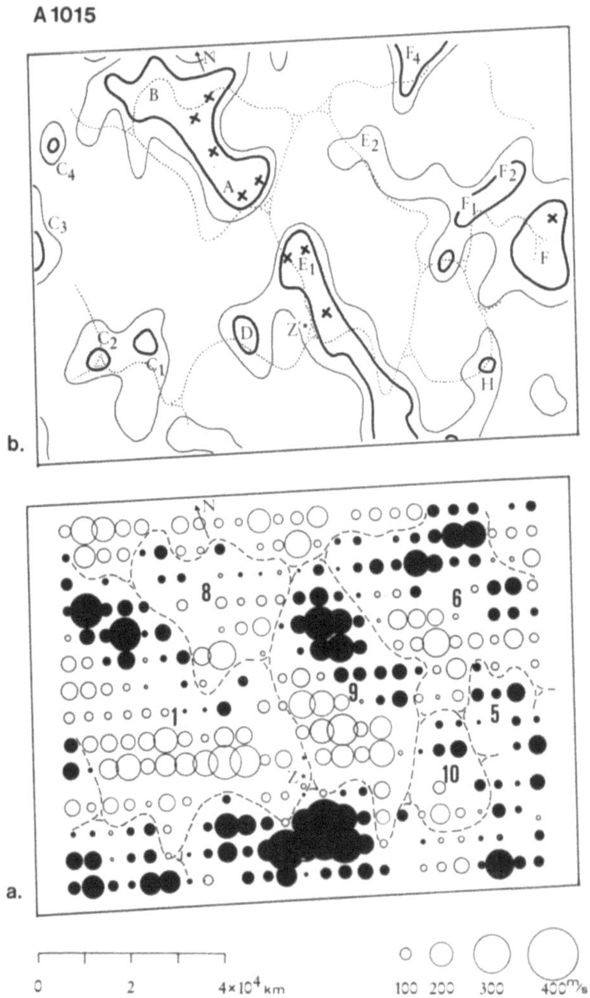

Fig. 5a and b. The complete observation of the field A 1015 represented as in Fig. 4

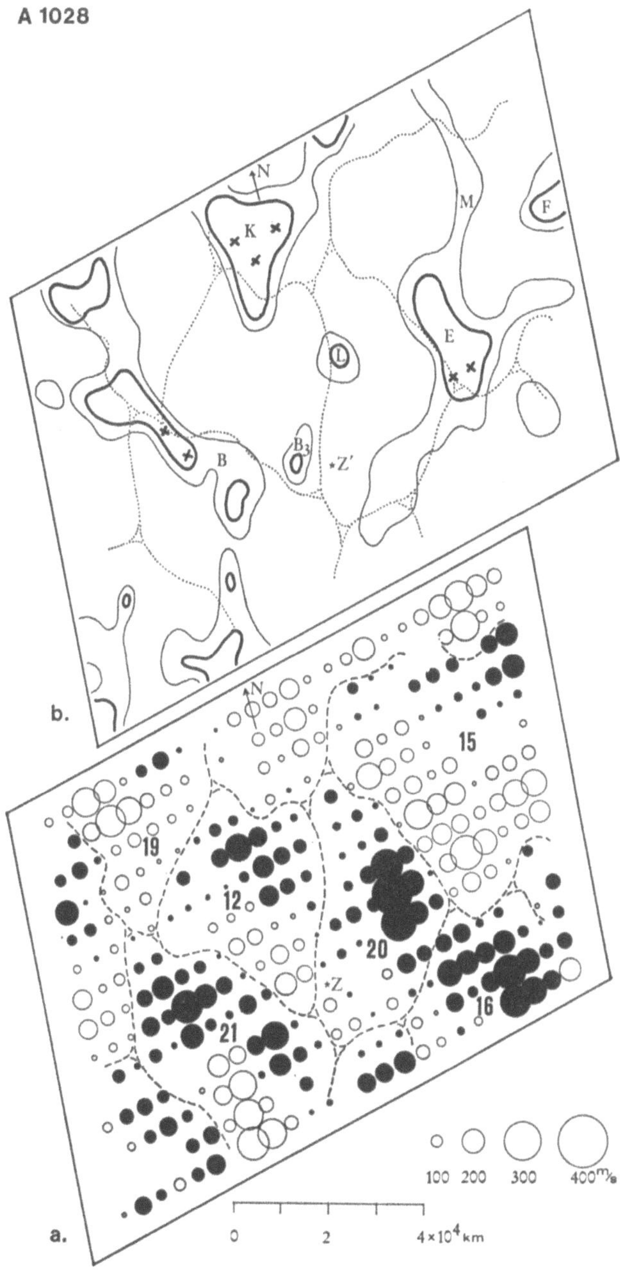

Fig. 6a and b. The complete observation of the field A 1028 represented as in Fig. 4. Note the Ca II emission "bridge" *M*

Conclusion

Although the work is still in progress, and some of the aspects of the problem have not been considered yet, few preliminary conclusions can be reached at now.

1) The effect of Coriolis acceleration is measurable on the supergranular scale, providing the velocity r.m.s. errors (mainly small-scale velocity noises) are not higher than ± 50 to ± 60 m/s.

2) The observed approaching velocity enhancements seem to fit into the picture of one-turn convective supergranular motion. As about four (more or less clear) such cases have been at all observed out of 50 identified cells, this stage of the evolution of a supergranule should be quite a short-living phase (almost eruptive). It can last, perhaps, not more than, or about, 10% of the life-time of a cell. (In the case of supergranule No. 43 it lasted certainly less than 6 hours.) However, this "eruptive" phase of approaching velocities may not be very typical, and this could also be the reason for the small number of the observed cases.

In any case, some room must be left in our speculations for any other more delicate form of convection, or for some contribution of oscillatory phenomena to the nature of the supergranular motion – the problems that have to be investigated yet.

3) As far as Ca II emission and supergranular velocity field interaction is concerned, some simple cases of emission network following the mass motion have been observed here. But, to understand some more complicated situations (e.g. the isophote *A*, Fig. 5), the necessary link between the two phenomena – the magnetic field – has to be observed as extensively as possibly.

Acknowledgements

The work has been started at the Kodaikanal Astrophysical Observatory during my tenure of a scholarship by the Indian Government, and completed as a regular program at the Astronomical Observatory in Belgrade.

I thank the Kodaikanal Observatory for their hospitality and the director Dr. M. K. V. BAPPU, for suggesting the problem and for the supervision of the first stage of the work. I am also grateful to Dr. M. VUKIČEVIĆ-KARABIN for a number of valuable discussions.

References

DEUBNER, F. L.: Solar Phys. **17**, 6 (1971).
FRAZIER, E. N.: Solar Phys. **14**, 89 (1970).
LEIGHTON, R. B., NOYES, R. W., SIMON, G. W.: Astrophys. J. **135**, 474 (1962).
MUSMAN, S., RUST, D. M.: Solar Phys. **13**, 261 (1970).
SIMON, G. W., LEIGHTON, R. B.: Astrophys. J. **140**, 1120 (1964).
TANENBAUM, A. S., WILCOX, J. M., FRAZIER, E. N., HOWARD, R.: Solar Phys. **9**, 328 (1969).

Discussion on the Solar Eclipse of 1973

By M. Rigutti
Osservatorio Astronomico Capodimonte, Napoli, Italy

This session took the form of an open discussion on the subject after an introduction by M. Rigutti who summarized what has already been done in trying to get the best organization for the observation of the June 1973 total solar eclipse.

Thanks to the efforts and interest of Dr. E. R. Dyer and Prof. C. de Jager two international meetings have already been organized for discussing problems connected with the observation of the eclipse, the first in Seattle, U.S.A., during the 15th Annual Meeting of the COSPAR, and the second in Madrid, Spain, during the 16th Annual Meeting of the COSPAR. A number of prospective observers attended these meetings and practical and scientific information have been shared among the participants.

Two issues of the Solar Eclipse 1973 Bulletin have been published and distributed by the U.S.A. eclipse coordinator Mr. R. R. La Count. The first, besides general information, made available to all interested persons the following five reports:

1) Report on Site Inspection in Mauritania by M. Waldmeier.

2) Prospecting Mission for the Site of the Eclipse by B. Fort.

3) Site Survey Report for the African Eclipse 1973 Expedition by Members of the U.S.A. National Center for Atmospheric Research.

4) NASA Site Inspection Report-Cape Verde Islands and Mauritania by James W. Gray.

5) Meteorological Planning Study for the 1973 African Eclipse by Neldur Medrud Jr.

6) U.S. Naval Observatory Circular N. 135, Total Eclipse of 30 June 1973 by Julena S. Duncombe.

The first gives information about Mauritania.

The second gives information about Mauritania and Niger.

The third gives information about Mauritania, Mali and Niger.

The fourth gives information about Mauritania and Cape-Verde Islands.

The fifth gives a number of meteorological informations along the eclipse path.

The sixth is the circular of the U.S.N.O. No. 135 prepared by Mrs. J. S. Duncombe and distributed at the end of 1972.

The second issue of the Bulletin included 44 observational programs which arrived to Mr. La Count as responses to the Solar Eclipse Questionnaire distributed together with the first issue of the Solar Bulletin. Besides, the second issue of the Bulletin made available further information:

1) An "Account of a Visit to Niger" by J. F. JAMES.

2) Report on Capacity of the Nouakchott Wharf from the American Embassy in Nouakchott, Mauritania.

3) Excerpts from the Solar Eclipse Bulletin of the Sudan National Council for Research.

Finally some interesting letters and the minutes of earlier meetings on the 1973 eclipse were given in the same Bulletin.

At the Madrid meeting Mme. ABDALLAHI and Dr. AGIB were also present on behalf of Mauritania and the Sudan, respectively. What they reported and their comments and suggestions can be found in the second issue of the Solar Eclipse Bulletin. Also in Madrid, Mrs. P. OWTRAM was present on behalf of the B.B.C. She was there because the European Broadcasting Union is interested to organize a live television coverage of the eclipse from somewhere in Africa.

Finally, it has been planned to get at least two IAU representatives willing to go to Africa to contact directly and officially the African Authorities on behalf of the IAU for getting from them as much assistance as possible for the various observing parties, as well as for the amateurs and television people who will crowd some localities of Mauritania and, of perhaps, Niger. For this purpose, the General Secretary of the IAU, Prof. C. DE JAGER, and the Chairman of the Working Group on Solar Eclipses, Prof. M. RIGUTTI, might be the most appropriate persons.

The occasion of the IAU regional meeting in Athens seemed to be a very good one for discussing a little longer some problems connected with the 1973 eclipse. Of course, the situation is already sufficiently clear: the eclipse will be very difficult and several parties will meet serious logistic problems, but all that can be done to facilitate the work of the parties should be done in order to increase the probability of success.

The discussion among the persons present at the meeting has been very interesting and stimulating. The contributions and remarks by M. K. M. ALY, M. K. V. BAPPU, R. FALCIANI, J. HOUTGAST, C. DE JAGER, D. H. MENZEL, M. RIGUTTI, and others led to the discussion of further details of the general and the particular aspects of the international organization and coordination of the eclipse activities.

From this discussion it resulted that the great majority of professional and/or amateur observers will be concentrated in Mauritania, Niger and, possibly, Malì. To these countries the action of the two IAU representatives should then be primarily devoted and they should contact their political and administrative Authorities for discussing with them the requirements of all persons engaged with the eclipse observations. Moreover the two IAU representatives should consider with these Authorities the possible observing sites and discuss with them all practical aspects involved. For this purpose they should make an actual site inspection in those at least sites of Mauritania, Malì and Niger where it is expected that most observers will go.

A very important point has been raised regarding the need of a central and really effective center of coordination. After a thorough discussion it has been accepted that only the National Scientific Foundation of the U.S.A. might be in the position of acting as such a coordination center. For this, it would be sufficient for the N.S.F., to extend to all astronomers the services and the organization already provided for the U.S.A. astronomers. This extension should not represent a very heavy task because the N.S.F. is already collecting various information and makes it available to all interested. For this reason Prof. C. DE JAGER promised to contact Mr. R. LA COUNT in order to investigate with him the possibility of having the N.S.F. as a super-coordination office for the 1973 eclipse.

Franck-Condon Factors for the Interpretation of the U.V. Spectrum Observed by Mariners IV, V, VI, VII, in the Atmospheres of Mars and Venus.

By R. Botter
D.R.A.-SRIRMA, C.E.N. Saclay B.P. N°2, 91190 Gif sur Yvette, France
and B. Petropoulos
Département des Recherches Physiques, Laboratoire Associé au C.N.R.S. No. 71, Université de Paris, Paris, France
With 2 Figures

Abstract

Franck-Condon factors have been calculated for the principal electronic transitions that have been observed with the spectrometers of Mariners in the U.V. spectrum of Mars and Venus. The R.K.R. Cooley method was used for this calculations.

Introduction

The strongest band systems observed on Mars are those due to the $B\Sigma_u^- - \tilde{X}^2\Pi_g$ and $\tilde{A}^2\Pi_u - \tilde{X}^2\Pi_g$ transitions of CO_2^+ and the $A^3\Pi - X^1\Sigma^+$ transition of CO[(1)]. The emission spectrum of CO_2 excited by electron impact can be divided into three different wave-length regions

1) – The 2800 Å–4500 Å interval corresponding to the CO_2^+ emission region.

2) – The 1950 Å–2500 Å interval in which the first negative system and the Cameron bands of CO are observed: $CO^+(B^2\Sigma^+ \rightarrow X^2\Sigma^+)$, $CO(A^3\Pi \rightarrow X^1\Sigma^+)$.

3) – The 1250 Å–1950 Å interval in which the atomic emissions predominate.

Emission cross-sections have recently been measured by Ajello[(2)] for various bands of CO_2^+, CO^+ and CO obtained by electron impact of CO_2. The electronic transitions observed by this author are given in Tables 1 and 1a.

Table 1. Observed transitions by Ajello (1972) for the ionisation of CO_2

1) 2800–4500 Å $CO_2(X^1\Sigma_g^+)+e \rightarrow CO_2^+(B^2\Sigma_u^+)+2e$ $CO_2^+(\tilde{B}^2\Sigma_u^+) \rightarrow CO_2^+(\tilde{X}^2\Sigma_g)+\Sigma h\nu_{v'v''}$ $CO_2(X^1\Sigma_g^+)+e \rightarrow CO_2^+(A^2\Pi_u)+2e$ $CO_2^+(A^2\Pi_u) \rightarrow CO_2^+(X^2\Pi_g)+\Sigma h\nu_{v'v''}$	
2) 1950–2500 Å	3. $CO_2(\tilde{X}^1\Sigma_g^+)+e \rightarrow CO(a^3\Pi)+O$ $CO(a^3\Pi) \rightarrow CO(\tilde{X}^1\Sigma^+)$
1. $CO_2(\tilde{X}^1\Sigma_g^+)+e \rightarrow CO^+(\tilde{B}^2\Sigma_u^+)+O+2e$ $CO^+(\tilde{B}^2\Sigma_u^+) \rightarrow CO^+(X^2\Sigma^+)+\Sigma h\nu_{v'v''}$	4. $CO_2(\tilde{X}^1\Sigma_g^+)+e \rightarrow CO(A^1\Pi)+O+e$ $CO(A^1\Pi) \rightarrow CO(X^1\Sigma^+)+\Sigma h\nu_{v'v''}$
2. $CO_2(\tilde{X}^1\Sigma_g^+)+e \rightarrow CO_2^+ + e$ $CO_2^+ \rightarrow CO(a^3\Pi)+O$ $CO(a^3\Pi) \rightarrow CO(X^1\Sigma^+)+\Sigma h\nu_{v'v''}$	5. $CO_2(X^1\Sigma_g^+)+e \rightarrow CO^+(A^2\Pi)+O+e$ $CO^+(A^2\Pi) \rightarrow CO^+(B^2\Sigma^+)$

Table 1a. Observed transitions by AJELLO (1972) for the ionisation of CO

$CO(X^1\Sigma^+)+e\rightarrow CO^+(A^2\Pi)+2e$ $CO^+(A^2\Pi)\rightarrow CO(X^1\Sigma^+)+\Sigma h\nu_{v'v''}$	$CO(X^1\Sigma^+)+e\rightarrow CO^+(B^2\Sigma^+)+2e$ $CO^+(B^2\Sigma^+)\rightarrow CO^+(X^2\Sigma^+)$
$CO(X^1\Sigma^+)+e\rightarrow CO^+(A^2\Pi)+2e$ $CO^+(A^2\Pi)\rightarrow CO^+(B^2\Sigma^+)$	$CO(X^1\Sigma^+)+e\rightarrow CO^+(A^1\Pi)+2e$ $CO^+(A^1\pi)\rightarrow CO(X^1\Sigma^+)$

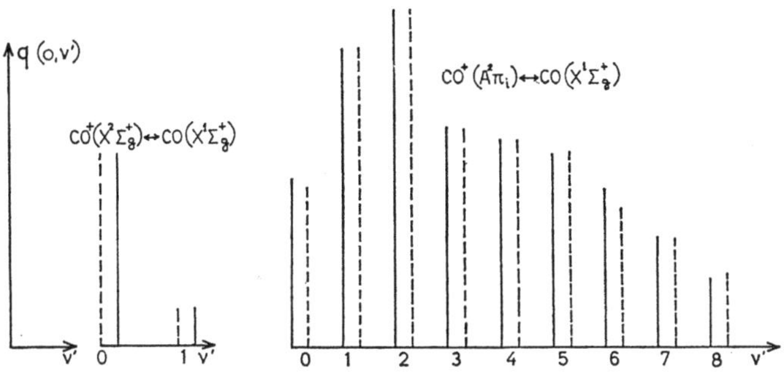

Fig. 1. Comparison of the experimental and calculated Franck-Condon factors for the $O_2(X^3\Sigma_g^-)\rightarrow O_2^+(b^4\Sigma_g^-)$ transition. ——— Calculated in this work. ------ Measured by TURNER, D. W.

In order to reproduce the spectra obtained by Mariners IV, V, VI and VII, Franck-Condon factors and centroids for the $CO_2^+(A^2\Pi_u-X^2\Pi_g)$, $CO_2^+(B^2\Sigma_u^+-X^2\Pi_g)$ and $CO_2(X^1\Sigma_g)\rightarrow CO_2^+(\tilde{A}^2\Pi_u, B^2\Sigma_u)$ transitions have been calculated by PETROPOULOS[3] and by PETROPOULOS and BOTTER[4].

In this work calculations of Franck-Condon factors for the remaining electronic transitions in Table 1 and 1a are presented. Other calculations were also performed for O_2 and O_2^+ which have been shown to be present in the Martian atmosphere.

Franck-Condon Factors for the Ionization of CO and O_2 and for the Cameron Bands of CO

The Rydberg-Klein-Rees method[5] was used to determine the potential energy and the R_{min} and R_{max} of the various electronic states of CO and CO^+. Rees' functions were integrated using JARMAIN's procedure[6]. Comparison of the R_{min} and R_{max} with the values obtained using the method of VANDERSLISSE[7] show that these functions are independent of the integration procedure.

These values of R_{min} and R_{max} were used together with NUMEROV's procedure[8] in order to calculate the radial wavefunction, the Franck-Condon factors and centroids for the first, second and third electronic transitions of CO given in Table 1.

In Table 2, the Franck-Condon factors and the centroids are given for the $CO^+(A^2\Pi_u(v'')\leftarrow CO(X^1\Sigma_g(v'))$ transition.

In Table 3 the vibrational transition probabilities for the ionization of CO $(CO(X^1\Sigma_g)_{v'=0}\rightarrow CO^+(X^2\Sigma)_{v''})$ are listed, together with the results of

Table 2. Franck-Condon factors and centroides for the transition $CO^+ (A^2 \Pi_u) v'' \leftarrow CO(X^1 \Sigma_g^+) v'$ using the R.K.R. Cooley Method

v''	$v'=0$	$v'=1$	$v'=2$	$v'=3$
0	9.2274−001	7.3157−002	3.8391−003	2.2976−004
1	7.4558−002	7.6659−001	1.4452−001	1.3280−002
2	2.6178−003	1.5224−001	8.1634−001	1.9923−001
3	7.9691−005	7.8215−003	2.1608−001	4.8890−001
4	1.2818−007	1.8717−004	1.6584−002	2.6964−001
5	9.6002−007	2.2383−006	5.2736−004	2.7395−002
6	9.8203−007	8.3880−008	4.1592−006	1.0755−008
7	6.0458−008	5.2783−008	4.1406−009	3.7604−006
8	4.6047−007	1.2186−008	5.8201−008	1.2900−006
R Centroide Å				
0	1.1107+000	1.0008+000	8.8484−001	8.0395−001
1	1.2297+000	1.1191+000	1.0134+000	9.1125−001
2	1.3369+000	1.2324+000	1.1269+000	1.0225+000
3	1.3586+000	1.3500+000	1.2393+000	1.1352+000
4	1.5639−001	1.4521+000	1.3520+000	1.2433+000
5	1.1125+000	1.5540+000	1.4565+000	1.3590+000
6	1.0893+000	8.5133−001	1.7196+000	1.4607+000
7	1.4788+000	1.0663+000	8.3095−001	1.9969+000
8	1.1190+000	8.9956−001	1.3440+000	1.3312+000
R^2 *Centroide*				
0	1.2346+000	9.8991−001	7.5232−001	6.0773−001
1	1.4997+000	1.2557+000	1.0178+000	8.0749−001
2	1.7687+000	1.5085+000	1.2760+000	1.0380+000
3	1.8514+000	1.8017+000	1.5265+000	1.2978+000
4		2.0879+000	1.8068+000	1.5453+000
5	1.2229+000	2.3970+000	2.0996+000	1.8288+000
6	1.1863+000	5.4416−001	2.8335+000	2.1147+000
7	2.0441+000	1.1351+000		3.5903+000
8	1.2519+000	7.3715−001	1.8301+000	1.7072+000

Table 3. Franck-Condon factors for the transition $CO^+ (X^2 \Sigma_g) \leftarrow CO(X^1 \Sigma_g)$

V'	0	
V''	R.K.R.	WACKS [11]
0	0.9652 *E*-00	0.963 *E*-00
1	0.3446 *E*-01	0.363 *E*-01
2	0.3196 *E*-03	0.111 *E*-03
3	0.1507 *E*-06	0.0 *E*-06
4	0.1018 *E*-04	0.0 *E*-04
5	0.1583 *E*-06	0.0 *E*-06
6	0.1974 *E*-05	
7	0.1134 *E*-05	
8	0.1853 *E*-07	

WACKS[11] who has used a Morse potential in his calculation. The agreement between the two sets of results is good for the first two vibrational levels and only poor for the third.

Table 4. Comparison of the calculated and measured Franck-Condon factors for the transition $O_2^+(b^4\Sigma_g^-) \leftarrow O_2(X^3\Sigma_g^-)$

v'	Measured values	Calculated values of Franck-Condon factors	
	[10]	[11]	This work
0	20	20	20
1	19	17	16.5
2	11.9	9	8
3	7.12	3.4	3.1
4	3.56	1.3	1.1
5	1.1		
6	0.11		

[10] TURNER and MAY (1967) ref. [14]
[11] WACKS (1964) ref. [16]

Table 5. Comparison of the calculated and measured Franck-Condon factors for the transition $O_2(X^3\Sigma_g^-) \leftarrow O_2^+(A^2\Pi_u)$

v'	Measured intensity	Calculated values of Franck-Condon factors	
	[4]	[11]	Our values
0	—	0.15	0.14
1	—	0.65	0.60
2	1.2	1.5	1.5
3	3.0	2.6	2.6
4	3.9	3.6	3.7
5	4.5	4.4	4.5
6	4.7	4.7	5.0
7	4.8	4.8	5.0
8	4.4	4.6	4.8
9	4.1	4.1	4.3
10	3.6	3.6	
11	3.2	3.1	
12	2.5	2.6	

The $G(v)$ and $B(v)$ calculated by NUMEROV's procedure are in good agreement with the experimental values. This is a good test of the validity of the method of NUMEROV and we can conclude that it can be used successfully for the calculation of the Franck-Condon factors for the former transitions in CO.

U.V. spectroscopy has also established the existence of O_2 and O_2^+ in the Martian atmosphere[(9)]. For this reason, the Franck-Condon factors of the principal transitions for the ionization of O_2 ($O_2(X^3\Sigma_g^-) \rightarrow O_2^+(B^4\Sigma_g^-$ and $A^2\Pi_u$)) were calculated using the RKR procedure. These are shown in Fig. 1 and in Tables 4 and 5.

Comparison of the Calculated and Experimental Franck-Condon Factors

In Fig. 2, the calculated vibrational transition probabilities for the ionization of CO into the ground and the first excited states of the ion, and the experimental results of TURNER obtained by photoelectron spectroscopy[10] are compared. The agreement is generally good except for the 0 vibrational level of the $A^2\Pi$ state.

The same comparison is made in Fig. 1 for the $O_2(X^3\Sigma_g^-)_{v''=0} \rightarrow O_2^+(b^4\Sigma_g^-)$ transition. These values are also given in Table 4 together with the calculated results of WACKS[11] who has used a Morse potential in his calculations. The agreement between calculated and experimental results is not good, especially for the higher vibrational levels.

In Table 5, the experimental Franck-Condon factors of TURNER[10] for the $O_2(X^3\Sigma_g^-)_{v''=0} \rightarrow O_2^+(^4\Pi_u)$ transition are compared to the factors calculated by WACKS and to our own results. The agreement in this case is good even for the higher vibrational levels. The differences between experimental and calculated results are not significant.

There is a difference between the two sets of calculated values for the two transitions of O_2 (see Table 4 and 5).

In conclusion, reproduction of the spectrum of the Cameron bands observed by Mariners VI and VII can be obtained using the calculated Franck-Condon factors and the experimental results of AJELLO. This reproduction should be more precise for the high vibrational levels with Franck-Condon factors calculated by RKR procedure than with those calculated with a Morse potential because this latter method should not give good results for the high vibrational levels.

The calculated Franck-Condon factors can be used for the construction of a model in the Mars atmosphere[12].

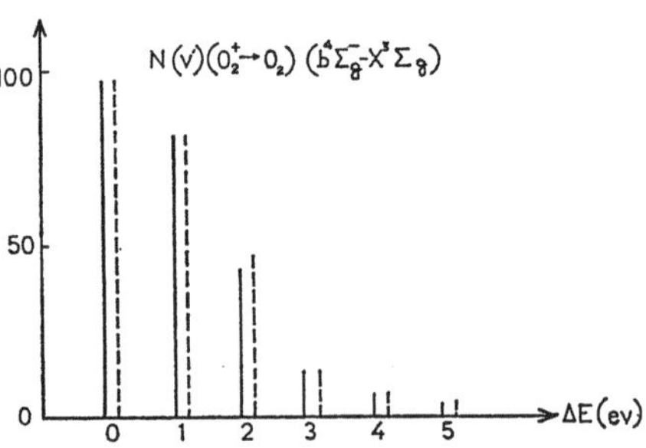

Fig. 2. Comparison of the Franck-Condon factors for the ionization of CO. ------ Measured by TURNER, D. W. —— Calculated in this work

References

1. BARTH, C. A.: Appl. Opt. **8**, 1295 (1969).
   BARTH, C. A., HORD, C. W., PEARCE, J. B., KELLY, K. K., ANDERSON, G. P., STEWART, A. I.: Mariner 1971 6 and 7 UV spectrometer experiment. J. Geophys. Res. **76**, 2213 (1971).
2. AJELLO, J.: J. Chem. Phys. **53**, 3160–3170 (1972).
3. PETROPOULOS, B.: Compt. Rend. Acad. Sci. **266**, 276 (1968).

4. Petropoulos, B., Botter, R.: Compt. Rend. Acad. Sci. **266**, 404 (1968).
5. Rydberg, R.: Z. Physic. **73**, 376 (1931).
   Klein, O.: Z. Physic. **76**, 226 (1932).
   Rees, A. L. G.: Proc. Roy. Soc. London A **59**, 998 (1947).
6. Jarmain, R. W.: Can. J. Phys. **38**, 216 (1960).
7. Vanderslisse, J. T., Davis, R., Weissman, S.: J. Chem. Phys. **43**, 1075 (1965).
   Gilmore, F. R.: Memorandum R. M., 1964, 4034 P. R.
8. Numerov, B.: Publ. Obs. Centr. Astr. Russ. **2**, 188 (1933).
9. Anderson, O. M.: Icarus **16**, 111 (1972).
10. Turner, D. W., May, P. D.: J. Chem. Phys. **46**, 1156 (1967).
11. Wacks, M. E.: J. Chem. Phys. **41**, 930 (1964).
12. Macris, C., Petropoulos, B.: In: J. Xanthakis (ed.), Solar activity and related interplanetary and terrestrial phenomena. Berlin-Heidelberg-New York: Springer 1973.

A New Model for the Atmosphere of Mars

By C. J. MACRIS
Research Center for Astronomy and Applied Mathematics, Academy of Athens, Athens, Greece
B. C. PETROPOULOS
Département des Rechèrches Physiques, Laboratoire Associé au C.N.R.S., No. 71 Université de Paris, Paris, France

With 3 Figures

Abstract

The authors used the computed Franck-Condon factors and the data secured by Mariner 6 and Mariner 7 flights; in this paper they suggest a new model of the martian atmosphere based on their calculations.

Our knowledge of the atmosphere of Mars and Venus has been completed in recent years through earth based IR and UV measurements and with the results of IR and UV spectrometer data furnished by Mariners 4, 6 and 7, for the martian atmosphere. Fig. 1 shows UV spectra measured by Mariners 6 and 7, and Table 1 contains the upper limits of abundances of the respective compounds. These have been estimated for the atmosphere of Mars by means of UV spectroscopy.

Before the Mariner explorations a high abundance of N_2 had been assumed for the atmosphere of Mars. The results obtained by Mariners however prove that

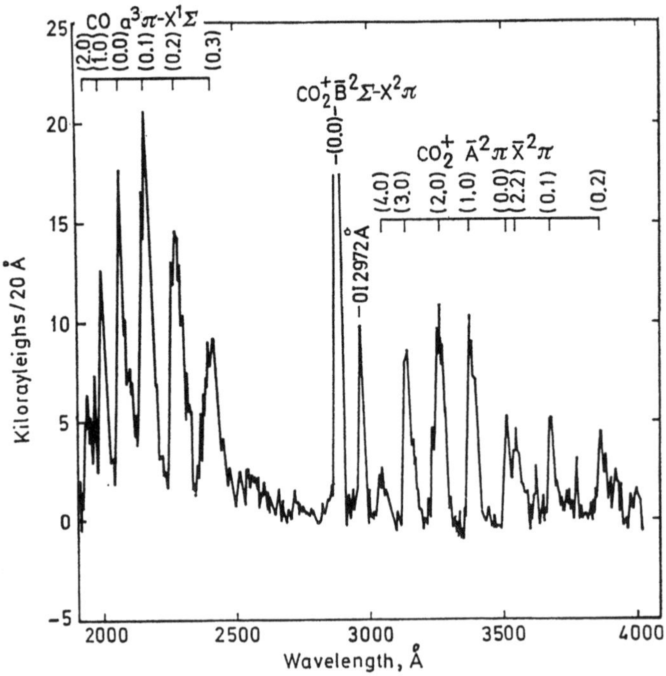

Fig. 1. Typical spectrum of the Martian Limb. (NASA SP-225, 1969, 99)

N_2 abundance does not exceed 1%, and CO_2 and CO_2^+ have been found to be more abundant than expected in the martian atmosphere and ionosphere.

The abundance of each compound (n relative) may be determined by using the Franck-Condon factors calculated by PETROPOULOS (1968), PETROPOULOS and BOTTER (1968) and BOTTER and PETROPOULOS (1972). The following relations have been used for these calculations:

$$n < \frac{4\pi I}{g(2\pi R_0 H)^{1/2}} \tag{1}$$

where the factor $(2\pi R_0 H)^{1/2}$ is the first term of the Chapman function. I = intensity, measured by Mariner 6 (BARTH *et al.*, 1971). R_0 = the distance to the centre of the Planet Mars. H = the scale of the height (20 km).

$$g = f(v', v'') \cdot g_{\text{mes}}(v', v''), \tag{2}$$

g is the scattering resonance factor and,

$$f(v', v'') = \frac{q(v', v'')}{\sum_{v'} q(v', v'')} \times \frac{q(v', v'')/\lambda^3(v', v'')}{\sum_{v''} \left[\frac{q(v', v'')}{\lambda^3(v', v'')}\right]}$$

where q is the Franck-Condon factor, $\lambda(v', v'')$ the wavelenght and $g_{\text{mes}}(v', v'')$ measured by Mariner (BARTH *et al.*, 1971).

We have used CHAMBERLAIN's model (1961) for calculating the abundances, and have made two working hypotheses: 1) That the CO_2 of the martian atmosphere is due to the action of the components of the solar flux (protons, electrons, photons, UV and X radiation). We have accordingly adopted a generalized CHAMBERLAIN's model and have taken into account all the possibilities of dissociation of the CO_2 molecule, as well as all possible electron transitions. 2) That the data supplied by Mariners 6 and 7 can be used for calculating the abundances function to the altitude for different vibrational states as shown in Table 2. The origin of the composition of the martian atmosphere in O_2, O_2^+, CO_2, CO_2^+, CO, CO^+, is correlated to the recently measured solar flux for Mars (HINTEREGGER, 1970). Fig. 2 shows the variations of the abundance (n relative). The upper limits

Table 1. Upper limits for the abundance of certain molecular species in Mars (BEER *et al.*, 1971)

Species	Principal frequency at which species was sought (cm^{-1})	Band	Type	Upper limit
HCl	2963.29	1−0	R3	11 μ-A
H_2S	2705.3	ν_1		30 mm-A
N_2O_4	2634	$\nu_1+\nu_{11}$		500 μ-A
NO_2	2914.75	$\nu_1+\nu_3$	QR7	250 μ-A
COS	2921.17	$\nu_1+\nu_3$		1.5 mm-A
C_3O_2	3084	$\nu_2+\nu_3$		200 μ-A
HCOH	2821.27	ν_5	P10	50 μ-A
HCOOH	2943	ν_2		70 μ-A
C_2H_2	2643.35	$\nu_1-\nu_5$	Q	20 mm-A

Table 2. Production of O and CO. The important sources of CO and atomic oxygen are:

1) Photodissociation

$$h\nu + CO_2 \rightarrow \begin{cases} CO(X^1\,\Sigma)+O(^3P) & \lambda\,2770\,\text{Å} \\ CO(X^1\,\Sigma)+O(^1D) & \lambda\,1670\,\text{Å} \\ CO(X^1\,\Sigma)+O(^1S) & \lambda\,1270\,\text{Å} \\ CO(\alpha^3\,\Pi)+O(^3P) & \lambda\,1080\,\text{Å} \end{cases}$$

2) Electron Dissociation

$$e^- + CO_2 \rightarrow \begin{cases} CO(X^1\,\Sigma)+O(^3P)+e^- \\ CO(\alpha^3\,\Pi)+O(^3P)+e^- \end{cases}$$

3) Ionospheric Recombination

$$CO_2^+ + e^- \rightarrow CO + O$$
$$CO_2^+ + O \rightarrow CO + O_2^+$$
$$O_2^+ + e^- \rightarrow O + O$$

Table 3. Models for the composition of the Martian atmosphere (ANDERSON *et al.*, 1972)

1. Nitrogen rich			2. Argon rich		
CO_2	80 m atm$^{\alpha}$	95.1%	CO_2	80 m atm$^{\alpha}$	66.6%
N_2	4	4.75%	A	40	33.3%
CO	0.056	666 ppm	Ne	0.08	666 ppm
A	0.048	571	CO	0.056	466
H_2O	0.04	475	H_2O	0.04	333
Ne	0.096×10^{-3}	1.1	Kr	0.0048	39.9
			Xe	0.0004	3.3

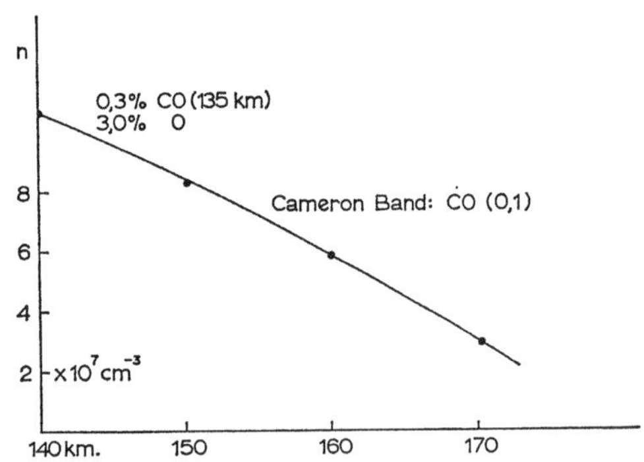

Fig. 2. Variation of the upper limit of the abundance of CO due to resonance scattering and calculated as a function of the altitude, in the Martian atmosphere. *n*: density due to resonance scattering of electrons

at the 135 km altitude of atomic oxygen (3%) and for CO (0.3%) in the atmosphere of Mars have been calculated and used in our present work. Variations of the band intensity of CO and O, as a function of the altitude have been measured by Mariners 6 and 7 (THOMAS, 1971).

For computing abundance we have used two different models: 1) In the first model we accepted a) the existence of 33% of Argon (ANDERSON *et al.*, 1972)

Table 4. Atmospheric parameters on Mars from Mariners 6 and 7 (HOGAN *et al.*, 1972)

Event	Surface temperature (°K)	Surface pressure (mb)	Peak altitude
Mariner 6 entry	254 ± 7	6.07 ± 0.36	136.5
Mariner 6 exit	152 ± 5	8.51 ± 0.48	—
Mariner 7 entry	221 ± 8	4.95 ± 0.24	137.5
Mariner 7 exit	209 ± 6	8.02 ± 0.45	—

Table 5. Temperature and molecular weight distribution (PITTS, 1968)

Geometrical altitude (km)	Temperature (°K)	Molecular weight
14	140.00	44.010
70	104.19	44.011
90	91.39	41.050
100	85.00	35.265
110	85.00	29.480
150	85.00	16.840
200	85.00	16.000
1000	85.00	16.000

(Table 3) and b) the non-existence of rares gases in the martian atmosphere. For both these cases we have taken the CO and O composition as is shown in Fig. 2 and Tables 6–9. 2) We have likewise considered the possibility that the reactions in the atmosphere of Mars and Venus produce $CO_2 \cdot CO_2^+$ with 50% $CO_2 \cdot CO_2^+$ and 50% CO_2 in the upper atmosphere of Mars. This model has been recently adopted by AIKIN (1972) for the atmosphere of Venus. Variation of the electron density may be interpreted on this basis for the atmospheres of both Mars and Venus.

By using this composition we may calculate the distribution of density, pressure, speed of sound, viscosity and temperature, function to altitude, in the atmosphere of Mars.

The program of PITTS (1968) has been adopted in an IBM 360 computer for these calculations.

The pressure and temperature data measured by Mariners 6 and 7, near the surface at different locations on Mars, tabulated in Table 4 (HOGAN, 1972) have been taken into account for our calculations; imput data of our program are shown in Tables 4 and 5. Having as basis the relations of the Pitts program we have computed the variations of the density, pressure, speed of sound, viscosity and temperature function to altitude in the atmosphere of Mars.

Tables 6–9 and Fig. 3 have been compared with the data computed by various authors (NASA SP-8010, 1968; RASOOL *et al.*, 1970; HUNTEN, 1972) who had adopted a different composition.

The values of density, pressure, speed of sound, viscosity and temperature for the atmosphere of Mars, calculated for the first time correctly on basis of the

Table 6. Altitude *vs* Density calculated for the Mars atmosphere. Alt. = 137.5 km, $T = 209^0$ K, $P = 8.02$ mb

Geometrical altitude (km)	a) 100% CO_2 b) 99.7% CO_2 0.3% CO c) 99% CO_2 1% CO	Aikin's model 50% CO_2 50% CO_2 · CO_2^+	Proton scattering 98% CO_2 2% CO	New model with argon 64.7% CO_2 33.3% A 2.0% CO	New model with argon 64% CO_2, 1% O_3, 30%A, 3% O, 2% CO	New model without argon 95.9% CO_2, 3% O, 0.1% CO, 1% O_3
0	2.03×10^{-5}	3.05×10^{-5}	2.02×10^{-5}	1.95×10^{-5}	1.92×10^{-5}	1.99×10^{-5}
20	2.66×10^{-6}	1.83×10^{-6}	2.68×10^{-6}	2.74×10^{-6}	2.77×10^{-6}	2.70×10^{-6}
40	1.45×10^{-7}	9.95×10^{-8}	1.46×10^{-7}	1.49×10^{-7}	1.51×10^{-7}	1.47×10^{-7}
60	5.95×10^{-9}	4.09×10^{-9}	5.98×10^{-9}	6.12×10^{-9}	6.19×10^{-9}	6.03×10^{-9}
80	1.72×10^{-10}	1.18×10^{-10}	1.73×10^{-10}	1.77×10^{-10}	1.79×10^{-10}	1.75×10^{-10}
100	3.90×10^{-12}	2.68×10^{-12}	3.92×10^{-12}	4.01×10^{-12}	4.05×10^{-12}	3.95×10^{-12}
120	1.45×10^{-13}	9.94×10^{-14}	1.45×10^{-13}	1.49×10^{-13}	1.51×10^{-13}	1.47×10^{-13}
140	1.13×10^{-14}	7.75×10^{-15}	1.13×10^{-14}	1.16×10^{-14}	1.17×10^{-14}	1.14×10^{-14}
160	1.70×10^{-15}	1.17×10^{-15}	1.71×10^{-15}	1.75×10^{-15}	1.77×10^{-15}	1.72×10^{-15}
180	3.41×10^{-16}	2.34×10^{-16}	3.43×10^{-16}	3.51×10^{-16}	3.55×10^{-16}	3.45×10^{-16}
200	7.18×10^{-17}	4.93×10^{-17}	7.21×10^{-17}	7.38×10^{-17}	7.46×10^{-17}	7.27×10^{-17}
220	1.59×10^{-17}	1.09×10^{-17}	1.60×10^{-17}	1.64×10^{-17}	1.66×10^{-17}	1.61×10^{-17}
240	3.59×10^{-18}	2.47×10^{-18}	3.61×10^{-18}	3.69×10^{-18}	3.74×10^{-18}	3.64×10^{-18}
260	8.25×10^{-19}	5.66×10^{-19}	8.28×10^{-19}	8.48×10^{-19}	8.58×10^{-19}	8.35×10^{-19}
280	1.92×10^{-19}	1.32×10^{-19}	1.93×10^{-19}	1.98×10^{-19}	2.00×10^{-19}	1.95×10^{-19}
200	4.55×10^{-20}	3.13×10^{-20}	4.57×10^{-20}	4.68×10^{-20}	4.73×10^{-20}	4.61×10^{-20}
320	1.10×10^{-20}	7.52×10^{-21}	1.10×10^{-20}	1.13×10^{-20}	1.14×10^{-20}	1.11×10^{-20}
340	2.68×10^{-21}	1.84×10^{-21}	2.69×10^{-21}	2.75×10^{-21}	2.78×19^{-21}	2.71×10^{-21}
360	6.64×10^{-22}	4.56×10^{-22}	6.67×10^{-22}	6.82×10^{-22}	6.90×10^{-22}	6.72×10^{-22}
380	1.67×10^{-22}	1.15×10^{-22}	1.68×10^{-22}	1.72×10^{-22}	1.74×10^{-22}	1.69×10^{-22}

Table 7. Altitude *vs* Pressure calculated for the Mars atmosphere. Alt. = 137.5 km, $T = 209^0$ K, $P = 8.02$ mb

Geometrical altitude (km)	a) 100% CO_2 b) 99.7% CO_2 0.3% CO c) 99% CO_2 1% CO	Aikin's model 50% CO_2 50% CO_2 · CO_2^+	Proton scattering 98% CO_2 2% CO	New model with argon 64.7% CO_2 33.3% A 2.0% CO	New model with argon 64% CO_2, 1% O_3, 30%A, 3% O, 2% CO	New model without argon 95.9% CO_2, 3% O, 0.1% CO, 1% O_3
0	8.02	8.02	8.02	8.02	8.02	8.02
20	6.85×10^{-1}	4.70×10^{-1}	6.88×10^{-1}	7.04×10^{-1}	7.12×10^{-1}	6.94×10^{-1}
40	3.38×10^{-2}	2.32×10^{-2}	3.39×10^{-2}	3.47×10^{-2}	3.51×10^{-2}	3.42×10^{-2}
60	1.24×10^{-3}	8.53×10^{-4}	1.25×10^{-3}	1.28×10^{-3}	1.29×10^{-3}	1.26×10^{-3}
80	3.29×10^{-5}	2.26×10^{-5}	3.31×10^{-5}	3.39×10^{-5}	3.43×10^{-5}	3.34×10^{-5}
100	7.81×10^{-7}	5.36×10^{-7}	7.85×10^{-7}	8.03×10^{-7}	8.12×10^{-7}	7.92×10^{-7}
120	3.89×10^{-8}	2.67×10^{-8}	3.91×10^{-8}	4.00×10^{-8}	4.05×10^{-8}	3.94×10^{-8}
140	3.99×10^{-9}	2.74×10^{-9}	4.01×10^{-9}	4.11×10^{-9}	4.15×10^{-9}	4.05×10^{-9}
160	7.22×10^{-10}	4.95×10^{-10}	7.25×10^{-10}	7.42×10^{-10}	7.51×10^{-10}	7.31×10^{-10}
180	1.48×10^{-10}	1.01×10^{-10}	1.48×10^{-10}	1.52×10^{-10}	1.53×10^{-10}	1.49×10^{-10}
200	3.17×10^{-11}	2.18×10^{-11}	3.18×10^{-11}	3.26×10^{-11}	3.30×10^{-11}	3.21×10^{-11}
220	7.04×10^{-12}	4.83×10^{-12}	7.07×10^{-12}	7.23×10^{-12}	7.32×10^{-12}	7.13×10^{-12}
240	1.59×10^{-12}	1.09×10^{-12}	1.59×10^{-12}	1.63×10^{-12}	1.65×10^{-12}	1.61×10^{-12}
260	3.64×10^{-13}	2.50×10^{-13}	3.66×10^{-13}	3.74×10^{-13}	3.79×10^{-13}	3.69×10^{-13}
280	8.49×10^{-14}	5.83×10^{-14}	8.53×10^{-14}	8.73×10^{-14}	8.83×10^{-14}	8.60×10^{-14}
300	2.01×10^{-14}	1.38×10^{-14}	2.02×10^{-14}	2.07×10^{-14}	2.09×10^{-14}	2.04×10^{-14}
320	4.84×10^{-15}	3.32×10^{-15}	4.86×10^{-15}	4.97×10^{-15}	5.03×10^{-15}	4.90×10^{-15}
340	1.18×10^{-15}	8.11×10^{-16}	1.19×10^{-15}	1.21×10^{-15}	1.23×10^{-15}	1.20×10^{-15}
360	2.93×10^{-16}	2.01×10^{-16}	2.94×10^{-16}	3.01×10^{-16}	3.05×10^{-16}	2.97×10^{-16}
380	7.38×10^{-17}	5.06×10^{-17}	7.41×10^{-17}	7.58×10^{-17}	7.67×10^{-17}	7.47×10^{-17}

Table 8. Altitude *vs* Speed of Sound (m/sec) for the Mars atmosphere. Alt. = 137.5 km, $T = 209^0$ K, $P = 8.02$ mb

Geo-metrical altitude (km)	a) 100 CO_2 b) 99.7% CO_2 0.3% CO c) 99% CO_2 1% CO	Aikin's model 50% CO_2 50% $CO_2 \cdot CO_2^+$	Proton scattering 98% CO_2 2% CO	New model with argon 64.7% CO_2 33.3% A 2.0% CO	New model with argon 64% CO_2, 1% O_3, 30%A, 3% O, 2% CO	New model without argon 95.9% CO_2, 3% O, 0.1% CO, 1% O_3
0	230	178	231	240	242	233
5	217	179	218	225	226	219
10	203	179	203	209	209	204
15	190	179	190	195	195	191
20	188	177	188	193	193	189
25	186	175	186	190	190	186
30	184	173	184	188	188	184
40	180	169	180	184	184	180
50	175	165	175	179	179	175
60	170	160	170	174	174	171
70	166	156	166	169	169	166
80	163	146	163	167	166	162
90	161	140	161	164	159	156
100	167	144	167	170	163	160
120	194	167	193	197	189	185
140	222	191	222	226	217	212
160	243	209	243	247	237	232
180	246	211	246	250	240	234
200	248	214	248	252	242	237
220	248	214	248	252	242	237

Table 9. Altitude *vs* Viscosity calculated for the Mars atmosphere. Alt. = 137.5 km, $T = 209^0$ K, $P = 8.02$ mb

Geo-metrical altitude (km)	a) 100 CO_2 b) 99.7% CO_2 0.3% CO c) 99% CO_2 1% CO	Aikin's model 50% CO_2 50% $CO_2 \cdot CO_2^+$	Proton scattering 98% CO_2 2% CO	New model with argon 64.7% CO_2 33.3% A 2.0% CO	New model with argon 64% CO_2, 1% O_3, 30%A, 3% O, 2% CO	New model without argon 95.9% CO_2, 3% O, 0.1% CO, 1% O_3
0	0.72×10^5	0.86×10^5	1.06×10^5	1.23×10^5	1.23×10^5	1.08×10^5
5	0.70	0.74	0.91	1.06	1.06	0.92
10	0.67	0.62	0.76	0.89	0.90	0.77
15	0.65	0.53	0.66	0.77	0.78	0.67
20	0.64	0.52	0.64	0.75	0.76	0.65
25	0.62	0.50	0.63	0.74	0.74	0.63
30	0.61	0.49	0.61	0.72	0.72	0.62
35	0.59	0.48	0.60	0.70	0.71	0.60
40	0.58	0.47	0.58	0.69	0.69	0.59
45	0.56	0.46	0.57	0.67	0.67	0.58
50	0.55	0.45	0.55	0.65	0.66	0.56
55	0.54	0.44	0.54	0.64	0.64	0.55
60	0.52	0.42	0.53	0.62	0.63	0.53
65	0.51	0.41	0.51	0.61	0.61	0.52
70	0.50	0.40	0.50	0.59	0.60	0.51
75	0.48	0.39	0.49	0.58	0.58	0.49
80	0.48	0.39	0.48	0.57	0.57	0.48
85	0.47	0.38	0.47	0.56	0.56	0.48
90	0.46	0.37	0.46	0.55	0.55	0.47
95	0.45	0.37	0.45	0.54	0.54	0.46
100	0.44	0.36	0.45	0.53	0.53	0.45

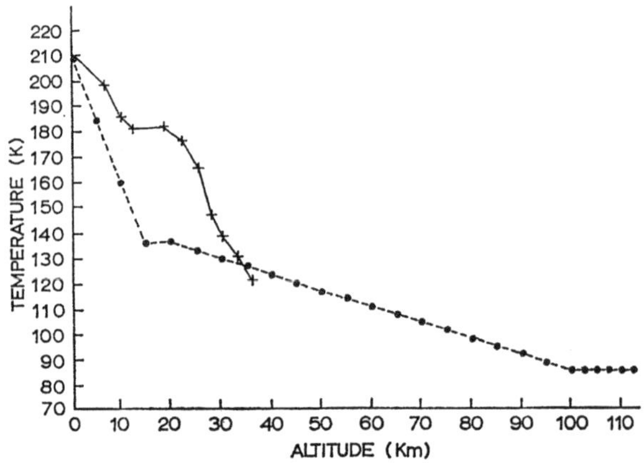

Fig. 3. Temperature *vs* altitude for Mariner 7 exit. Continuous line: Curve 1 of the Fig. 28 given by HOGAN *et al.* (NASA TN D-6683, p. 30) for 100% CO_2. Dashed line: Curve MACRIS-PETROPOULOS, for 95.9% CO_2, 3% O, 1% O_3, and 0.1% CO

abundance are given in this paper. The different possibilities of decomposition of CO_2 by the solar flux have been examined. These results are available in the case of Mars for the lower and the upper atmosphere of the planet. As concerns the middle atmosphere we may assume that it consists of 50% of the complex ion $CO_2 \cdot CO_2^+$ and 50% of CO_2.

We do not know yet exactly the values of the pressure and temperature on the surface of Mars. Estimation of the variation of these quantities will be possible when the Viking project has furnished the necessary data.

From the results obtained we may draw the conclusion that the new models very much depend on the abundances of the CO_2 fragments, i.e. CO and O; and the percentage of these fragments and their importance in the composition of the atmosphere depends on the intensity of the solar flux which varies depending on the time of observation, the position of Mars, the solar events, etc.

If we take into consideration the scattering by electrons and the UV absorption by CO_2, we find that the abundances of CO and O at a height of 135 km are, in the atmosphere of Mars, 0,3% and 3% respectively. Fig. 3 shows the variation of temperature with height based on these percentages. This temperature variation differs from the corresponding one to the Mars' atmosphere, the composition of which is 100% of CO_2.

Mariners 6 and 7 have measured at various points of the martian atmosphere different values of proton flux (STEWART, 1971) which justifies different abundance of CO and O through proton impact. In this paper we consider 2% as an upper limit for the abundance of CO, and we calculated accordingly the density, pressure, etc, function to altitude (Tables 6–9 and Fig. 3). If we do not take into consideration CO and O (CO_2=100%) these quantities differ by the new models calculated values.

We therefore offer a new model in which the variations of the proton flux and of proton scattering are taken into consideration to explain the dissociation of CO_2.

If we accept as valid, also in the case of Mars, AIKIN's (1972) hypothesis that the atmosphere of Venus has a $CO_2 \cdot CO_2^+$ content amounting to 50% then the martian pressure, density, etc. differ appreciably compared to those computed for $CO_2 = 100\%$. The differences in these quantities may accordingly be attributed to the changes of electron density in the atmosphere of Mars.

If we assume the existence of Argon in the atmosphere of Mars, then the distribution of density and pressure (Tables 6, 7) is different.

The existence of O_3 in the atmosphere of Mars has been recently investigated by GRAMS (1972) who calculated that the upper limit of ozone content must be of the order of 1%. We have taken this quantity into consideration in the course of this work.

Acknowledgements

This research has been sponsored in part by the Empirikion Foundation; we express our thanks to the Board.

We likewise wish to thank Dr. CH. FRANGAKIS who offered valuable personal assistance in setting the program for the IBM 360 computer.

References

AIKIN, A. C.: Nature **235**, 10 (1972).
ANDERSON, D. M., BIERMANN, K., ORGEL, L. E., ORO, J., OWEN, T., SHULMAN, G. P., TOULMIN III, P., UREY, H. C.: Icarus **16**, 111 (1972).
BARTH, C. A., HORD, C. W., PEARCE, J. B., KELLY, K. K., ANDERSON, G. P., STEWART, A. I.: J. Geophys. Res. **76**, 2213 (1971).
BEER, R., NORTON, R. H., MARTONCHIK, J. V.: Icarus **15**, 1 (1971).
BOTTER, R., PETROPOULOS, B.: Related interplanetary and terrestrial phenomena, Berlin-Heidelberg-New York: Springer (to be published).
CHAMBERLAIN, J. W.: 1961, Physics of the aurora and airglow, p. 582. Acad. Press, New York.
GRAMS, S., GRAMS, G.: NASA, TN D-6812 (1972).
HINTEREGGER, H. E.: Ann. Geophys. **26**, 547 (1970).
HOGAN, J. S.: NASA TN D-6683, 29 (1972).
HUNTEN, D. M.: Physics of the solar system, NASA SP-300, 197 (1972).
NASA SP-8010, 17 (1968).
NASA SP-225, 99 (1969).
PETROPOULOS, B.: Compt. Rend. Acad. Sci., France **266**, 276 (1968).
PETROPOULOS, B., BOTTER, R.: Compt. Rend. Acad. Sci., France **266**, 404 (1968).
PITTS, D. E.: NASA TN D-4292 (1968).
RASOOL, S. I., HOGAN, J. S., STEWART, R. W.: J. Atmospheric Sci. **27**, 841 (1970).
STEWART, R. W.: J. Atmospheric Sci. **28**, 1069 (1971).
THOMAS, G. E.: J. Atmospheric Sci. **28**, 859 (1971).

Discussions

FALCIANI, R.:
How do you explain physically the plateau in the temperature distribution in your Mars atmosphere model at the height of about 20 km?

PETROPOULOS, B. C.:
The same plateau can be observed in the distribution of the temperature in the Earth's atmosphere and for this reason we change the composition of the atmosphere in this altitude. It is more accentuated if you accept a 100% CO_2 atmosphere as it has been admitted by HOGAN. HOGAN has also used a correction of the results of Mariners, on basis of the index of refraction of the Mariner's transmission in the Martian atmosphere. We have not used this correction. But it appears that change of the derivative of temperature produced this plateau, which is not very accentuated for our model.

KARABIN, M.:
The discontinuity of temperature in Martian atmosphere could be explained in similar way as is done in the Earth' upper atmosphere (because of change in composition)?

PETROPOULOS, B. C.:
The discontinuity of the temperature in Martian atmosphere cannot be explained in similar way as in the Earth' atmosphere, because the Martian atmosphere is more homogeneous than the Earth's atmosphere. We have calculated a different composition of O, CO and we have also discussed the existence of Ar and CO_2, CO_2^+ in the Martian atmosphere.

MALTBY, P.:
You have described a spherical symmetrical model. How large is the effect of the solar wind in producing an asymmetry?

PETROPOULOS, B. C.:
We have accepted that the effect of the solar wind is negligeble, in our calculation. A model has been proposed by CLOUTIER who has admitted a large effect of solar wind producing an asymmetry. Our calculation has been realised for vertical altitudes and for the different entry and exit of Mariners. You can see that in the table no notable difference exists in the pressure and temperature measured by Mariners. Consequently, no notable difference exists for the calculation of the above values, for the different explorations by Mariners.

The homogeneous atmosphere of Mars permitted us to accept only a small solar wind effect in this calculation.

RIGUTTI, M.:
May I ask you if the measurements made by Mariners during their entry in the Martian atmosphere gave results different from those obtained during the exit?

If yes, how much different were they?

PETROPOULOS, B. C.:
Some results have been shown for the exit of Mariner 7. We have calculated for all the entries and exits of Mariners the pressure, temperature, density, viscosity, speed of sound. These results will be published in another paper.

The results of calculations of entry and exit are not very different.

GERARD, J.:
The mechanism for production of $CO_2 \cdot CO_2^+$ cluster ions assumed by AIKIN on Venus seems irrelevant for Mars. Indeed, it is a three-body reaction and de-

pending on square of pressure. The pressure being much lower on Mars, one can expect that this reaction does not occur referring to reaction rate used by AIKIN.

PETROPOULOS, B. C.:
We do not know experimentaly the mechanism of production of $CO_2 \cdot CO_2^+$ cluster ions.

AIKIN has proposed the existence of this ion in the atmosphere of Venus, on basis of a three-body reaction. But $CO_2 \cdot CO_2^+$ have not stabilisated in laboratory experiences, using the proposed by AIKIN's method, or another experimental method.

In the Mars atmosphere we have a great abundance of CO_2^+. And we have made only the hypothesis of the existence of $CO_2 \cdot CO_2^+$ for studying the variation of the density in the upper atmosphere where the composition in CO_2 and CO_2^+ is not known. In this region different mechanisms of production can exist and the atmosphere is not as homogeneous as in the other regions, due to different reactions (ionospheric recombination etc.).

Photometric Behaviour of Saturn's Rings as a Function of the Saturnocentric Latitudes of the Earth and the Sun

By S. A. Pyykkö
University Oulu, Oulu, Finland

With 2 Figures

The surface brightness of Saturn's rings as a function of the phase angle has been studied by several authors since the 19th century. So the phase curve of Saturn's rings, especially that of the ring *B*, is comparatively well known. But the dependence of the surface brightness from the elevation angles of the Earth and the Sun has not been extensively studied. Perhaps the only noticeable measurements have been made by Camichel during 1943–1957. Since he published the original measurements without corrections for various systematic errors, his results are not sufficiently accurate for the theoretical investigation of Saturn's rings. After obtaining Camichel's plates our working group at Aarne Karjalainen Observatory attacked anew the problem of the behaviour of Saturn's rings. The results are seen in Fig. 1.

We have studied two alternative explanations for these curves. First we calculated the effect of multiple scattering but noticed that it cannot be responsible for the brightening of ring *B*. The generally assumed optical thickness $\tau = 1$ did not give the observed increase in brightness. This was achieved with $\tau \sim \infty$, but then the theoretical curve did not ressemble the observed behaviour of ring *B*. In Fig. 1 it would be an almost straight line connecting both ends of the observed curve. So we had to reject the possibility of multiple scattering.

The second explanation studied for the behaviour of the surface brightness of ring *B* is stratification. We were led to this conclusion after considerations of celestial mechanics. If we assume that the particles possess different masses and if their mutual collisions are almost elastic, the calculations led to a special distribution of orbits, in which the orbital inclinations are inversely proportional to

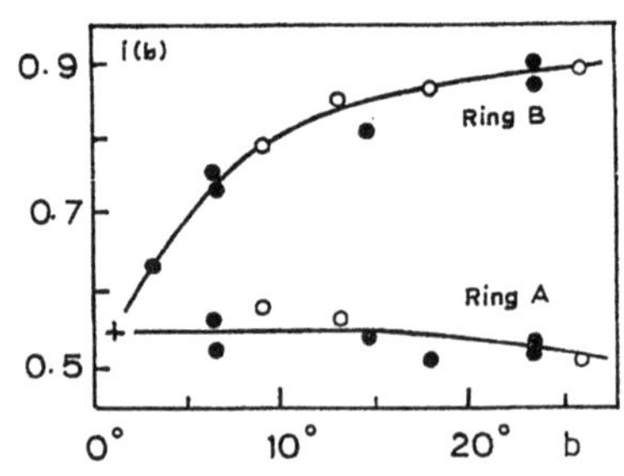

Fig. 1. The dots represent the measurements made essentially at constant phase 3° (a) 5. The circles represent the measurements with phase correction and the crosses the weighted mean of Focas' and Dolfus' data for the year 1966

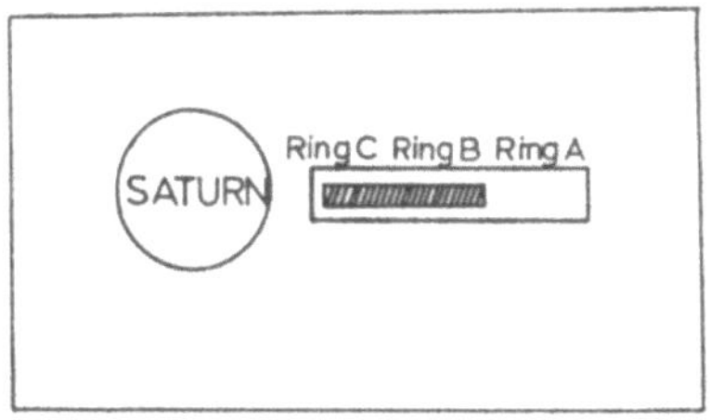

Fig. 2. Schematic cross-section of Saturn's rings

the square root of the mass. Consequently, the heaviest particles remain near the equatorial plane and the smaller particles form an envelope around it.

Fig. 2 is a schematic cross-section of Saturn's rings. The nucleus (shadowed area) is a mixture of large and small bodies, the envelope consists of small bodies only. The Cassini division is believed to be near the outer boundary of the nucleus.

As can be seen from Fig. 2 the path of the light ray will be different at different values of the elevation angles; when the elevation angles are large, the incident rays will be reflected mainly by the particles of the nucleus and at small angles the reflection shall take place in the envelope, whence the surface brightnesses of rings A and B are equal. The brightning of ring B can be understood if the larger particles have a higher albedo than the smaller ones. Our calculations showed that the heaviest particles must have a geometric albedo near unity and the envelope particles about 0.7. This difference can be understood by assuming that the larger bodies are condensations of the smaller and have a very complex surface structure. Multiple scattering in their surface layers is able to produce an increased value of the geometric albedo.

To improve the model given above we are investigating the opposition effect at different values of the solar and terrestrial elevation angles and the preliminary results seem to be in accordance with the stratified ring model.

The European Southern Observatory (ESO)

By A. Blaauw
European Southern Observatory, Hamburg, Germany

Abstract

The article summarizes the principal points of the agreement between the member states, the present status of the Organization and of its establishments, and the project of the 3.6 m telescope still to be completed. It describes the instrumentation now in operation on the mountain La Silla in the Chilean desert and the organization of their use by astronomers from institutes in the member states and by the semi-permanent staff of ESO, most of whom have their duty station in Chile. Current research projects are briefly reviewed as well as research prospects to be opened when the 3.6 m telescope becomes operational.

I. Basic Program and Membership

Member states of the European Southern Observatory (ESO) are Belgium, Denmark, France, the German Federal Republic, the Netherlands, and Sweden. The convention between these states was signed by the first (four)members in October 1962. It states that members should jointly establish an (optical) astronomical observatory for research in the southern hemisphere and describes the telescopes with which it is to be equipped: one of outstanding size (about 3 m), three of intermediate size, a large Schmidt type telescope and a telescope for the measurement of stellar positions. Auxiliary instrumentation for the telescopes is to be provided. The convention also specifies the procedure to be followed for the adoption of new member states.

II. Structure of ESO: Organizational and Geographical

The Observatory has been established in Chile, with the observatory proper in the Atacama desert, and office facilities in Santiago, A division for the design and construction of the large telescope is located in Geneva. The administrative headquarters including the Office of the General Director are in Hamburg.

The headquarters in Hamburg entertain the relations with the governments of the member states, and more particularly with the two delegates who represent these states in the ESO Council. For each state there are an astronomer and an administrative official. Council, in consultation with the General Director, establishes the financial basis on which the organization operates, it fixes the Rules and Regulations which define the status of the personnel, it supervises the general development of the organization within the policy decided on by the governments, etc. Present president of the ESO Council is Mr. A. ALLINE of the Department of Foreign Affairs in France. The Council is assisted by various advisory bodies: the Finance Committee (president C. ZELLE of Germany), the Instrumentation Committee (president J. BORGMAN of the Netherlands), and the Scientific Policy Committee (president B. STRÖMGREN of Denmark). This latter committee studies problems of long term development of the organization. For matters concerning the observing programs to be carried out with the existing telescopes, the Council and the ESO Directorate are advised by the Observing Programs Committee (OPC, president P. LEDOUX of Belgium).

The Hamburg office also administers the program of the visiting astronomers. Astronomers in the ESO countries can twice a year submit applications for observing time with the ESO telescopes. These applications are first submitted to, and discussed by, the OPC, which advises the ESO Directorate on the scientific merits. The Directorate next allocates time, taking into account not only these scientific aspects, but also instrumental and financial ones. To a certain degree, it also aims at adequate distribution over the nationalities of the member states.

The Hamburg office contains the principal administrative offices for financial matters, personnel matters, and for purchasing and shipments to Chile. In connection with the personnel services it is worth noticing that, contrary to "normal" state- or university institutes, ESO, due to its international status, has to provide its own system of personnel Rules and Regulations and continuously keep these adapted to modifying circumstances, like wage level trends.

The Telescope Project Division (TP-Div.) is located on the CERN grounds at Meyrin near Geneva. This division was created in September 1970 by agreement between CERN and ESO, for the purpose of the design, the construction and the erection of the 3.6 m telescope, and for the creation of the auxiliary instrumentation required for the initial operation of the telescope. Collaboration with CERN was sought for in order that ESO might profit from the extensive experience of CERN with regard to the realization of large instrumental projects and from the wide range of technical expertise present at CERN. Work at the TP-Division started already in October 1970. At the end of 1972 the first construction contracts (for the telescope building, the dome and the main structure of the telescope) are to be concluded.

III. Present Status

Except for the largest telescope, the program described in the ESO convention has now been realized: the most recent addition to the telescope park has been the Schmidt telescope. It is very similar to the Palomar 48-inch – the focal length and hence the scale are identical, but the aperture is 1 m instead of about 1.20 m. The unvignetted field accordingly is also somewhat smaller. This smaller aperture was chosen to allow a stronger mechanical design and thereby the mounting of heavy objective prisms. First successful plates have already been taken with the telescope, but before regular operation will be possible, the electronic drive system requires further adjustments.

Instruments in regular operation are:

a) The 1.5 m telescope, sometimes called the spectrographic telescope, because it is normally equipped with spectrographs. These are either a Cassegrain, or the coudé spectrograph.

b) The 1 m telescope, sometimes called the photometric telescope, because it normally carries photometric equipment.

c) The 50 cm telescope also used normally for photometric work.

d) A 40 cm double astrograph for objective prism radial velocity work of the type designed by FEHRENBACH; that is, one of the two telescopes carries a rotatable objective prism, whereas the other telescope serves for guiding. By means of measurements of the displacement between the first and the second spectral exposures taken with opposite direction of the dispersions, the Doppler-shift is measured.

In addition to these, ESO has access to two more telescopes:

a) A 50 cm telescope belonging to Copenhagen University, and

b) a 60 cm telescope belonging to Bochum University.

Of these two, one third of the time is available for ESO's staff and visiting astronomers.

As a rule, all these instruments are used in combination with the auxiliary apparatus provided by ESO. However, sometimes visiting astronomers bring their own specialized equipment after this has been adapted to attachment to our telescopes.

These observing facilities are located on the mountain La Silla in the Atacama desert, about 600 km north of Santiago. An area of about 600 km^2 around La Silla belongs to the observatory and protects the observing from interference by mining or other activities causing light or dust. 40 km of road had to be constructed to connect the mountain with the main highway leading from Santiago north along the southamerican coast.

Also on La Silla are established various facilities necessary for the observational operations: A hostel containing restraurant facilities, hotel rooms and limited office space, additional dormitories for lodging astronomical technical and administrative staff and the night assistants, power supply, water supply, supply stores, transport services, and workshops. Most of the personnel working on La Silla is of Chilean origin, about 20% are international staff mostly from the ESO member states.

Most of the administrative work in Chile is carried out in the office in Santiago, where also are located offices for the semi-permanent astronomical staff, electronic and mechanical technicians, etc. A large astronomical library, under construction, is also in this office building. On the grounds next to the office is a well equipped mechanical workshop, mostly used for the construction of new auxiliary instrumentation. A variety of measuring instruments, including a fully automatized Grant spectral measuring machine, are also in regular operation here. Most of the semi-permanent ESO international staff live with their families in the Santiago area and go for duty stays to La Silla. For the astronomical staff about one third of the observing time is available, with the remaining two thirds for the visiting astronomers. These latter, in between their observing sessions on La Silla, frequently use the measuring and library facilities in Santiago.

IV. The 3.6 m Telescope

This telescope, now in preparation by the TP-Division at Geneva, is expected to become operational in the course of 1976. Constructions for the building on the highest top of La Silla will start early 1973. The telescope has been designed basically as an all-purpose instrument, with possibilities for use at the primary, the Cassegrain, and the coudé focus, and a Ritchey-Chrétien optical system. In order to allow observations in the infrared, the entire top part of the telescope is interchangeable. Day time use for infrared work is foreseen. For coudé work a spacious laboratory, which may house several large spectrographs and experimental equipment, has been planned. Naturally, certain other services like photographic darkrooms, air conditioning, computer facilities, are also projected in the telescope building; yet, these have been kept to a minimum in order to avoid undesirable heat production and possible consequent turbulences and deterioration of the image quality. A number of additional services of these kinds, including electronic and optical laboratories, will be housed in a separate building, remote from the telescope.

Auxiliary equipment for the 3.6 m telescope is now being planned with the possible distinction between basic and specialized instrumentation. Whereas certain equipment will without doubt be required for future research, like the high-dispersion coudé spectrographs, and therefore may be considered as basic, other equipment will have to be developed in accordance with the needs following from the research projects. For the realization, priority will be given to the basic instrumentation. The development of the specialized instruments requires an early reflexion on the research projects to be carried out. Whereas it is very difficult to predict these 4 years ahead of time, it seems very likely that cosmological problems with emphasis on the observation of faint galaxies and quasi-stellar sources will deserve high priority. The same may be true for the observation of processes in interstellar matter.

V. Literature on ESO

Descriptions of the telescopes now in operation and of much of the auxiliary equipment may be found in the ESO Bulletins, which appear at irregular intervals and are sent to astronomical libraries all over the world. The project of the 3.6 m telescope has been described in the volume "ESO/CERN Conference On Large Telescope Design" (editor R. M. West), which is the proceedings of discussions held on this subject in March 1971. A conference "Auxiliary Instrumentation For Large Telescopes", held in May 1972, also under the auspices of ESO and CERN and published under this title (editors S. Laustsen and A. Reiz), contains basic material for the development of the telescope auxiliaries. In preparation for the programmes of the ESO Schmidt telescope, a conference was held in Hamburg in March 1972 of which the proceedings (editor U. Haug) are published under the title "The Role of Schmidt Telescopes in Astronomy".

Western European Cooperation in Space Astronomy

By C. de Jager
The Astronomical Institute, Utrecht, The Netherlands

With 5 Figures

Abstract

A review is given of the organizational structure of the ESRO (European Space Research Organization). We describe the main scientific programs in the field of astrophysics; we give a summary of the properties and scientific programs of the satellites launched so far, and show some of their results.

We also describe the scientific programs for the satellites yet in preparation, and attempt to give an outlook to the distant future.

I. The Structure and Functioning of ESRO

The ESRO's name (European Space Research Organization) shows, as those of organizations like CERN and ESO, some of the optimism of its founders, since it is actually only ten western European countries that have united their forces in ESRO. The organization, founded in 1964, has the following members: Belgium, Denmark, France, German Federal Republic, Italy, The Netherlands, Spain, Sweden, Switzerland and the United Kingdom of Great Brittain and Northern Ireland. Austria and Norway have the status of observers.

The ESRO's initial, exclusive purpose was to carry out scientific research by means of space vehicles, defined as rockets, satellites and space probes; balloons are not considered space vehicles.

A reorganization of the ESRO in 1971 has somewhat reduced the emphasis on scientific satellites and has added an "application program" centered around meteorological and communication satellites, and satellites for air-traffic control. The procedure for defining the scientific program is lengthy, and involves consultation of the European scientific community in as broad a measure as possible. There exists a *Launching Program Advisory Committee* (L.P.A.C.) of five persons, advised by two *Working Groups* (for the planetary system, and for general astrophysics) and a *Panel* for fundamental physics, each of them having about a dozen members. Furthermore small *ad hoc panels* are regularly established, to make detailed studies of certain specific missions, such as e.g. an excentric X-ray satellite using lunar occultations for positional determinations (HELOS), a geostationnary satellite for magnetospheric research (GEOS) ... etc. When, at the advice of the LPAC a specific future scientific mission has finally been accepted by the Council for inclusion in the scientific program, this mission is at that time only broadly defined. Then the scientific community in the member-states is invited to submit detailed proposals for the scientific payload of the relevant space vehicle.

With the growth of the European capacity and capability in space research the competition is usually very heavy; often institutes in different countries form *consortia* solliciting the inclusion of a certain experiment to the scientific payload. In this way a real international cooperation at a high scientific level between many institutes in the member countries has gradually grown, and also in that way ESRO has successfully contributed to intra-european cooperation.

One of the projects, now under discussion (the socalled H-MD-project) envisages even the cooperation of ESRO with NASA, which will lead to cooperation between institutes on either side of the Atlantic Ocean.

It is only obvious that large-scale cooperation of this kind must eventually lead to a relatively small number of institutes doing high-level scientific research, through a rather harsh, but highly efficient process of "survival of the fittest".

The main ESRO institutes are – besides the Headquarters in Paris –, the European Space Technology Center (ESTeC) in Noordwijk (the Netherlands), and the European Space Operations Center (ESOC) in Darmstadt (German Federal Republic).

The reorganization of 1971 brought the decision to concentrate the scientific effort on satellites. It was therefore agreed to terminate the acitivity in the field of plasma physics of the European Space Research Institute (ESRIN) in Frascati (Italy), and to abolish the ESRO rocket program, with the consequent closing of ESRANGE, the European rocket range near Kiruna (Nth. Sweden), which has since been taken over by the Swedish authorities.

II. ESRO Launchings

In the eight years since its founding six satellites have been launched, and about 170 rockets. About 40 to 50 scientific groups were involved in these experiments. The main characteristics of the satellites are given in Table 1. For further details reference is made to the Table, usually printed at the end of ESRO's Annual Report to COSPAR.

The *rocket* payloads were greatly varied in composition. Most of them contained equipment for upper atmospheric, ionospheric, and magnetospheric research, a few were directed to astrophysical research such as stellar photometry, solar X-ray spectrography, X-ray heliography or to energetic particle research ... etc.

Of the *satellites* launched so far, the ESRO I and II satellites are small and were launched by US Scout rockets; the highly excentric HEOS satellites form another class, and were both launched by a 3-stages-3-boosters Thor-Delta combination; the Thor-Delta 1 A satellite is the largest and most refined yet launched; it was brought into orbit by a Delta-N 2-stages, 3 boosters, combination.

Apart from the Boreas satellite, which had only an effective life time of a few weeks due to its very low orbit, the satellites have offered during a long time a wealth of interesting scientific results.

Table 1. Main properties of ESRO's satellites launched until 1972

Name	Main scientific program	Inclination (°)	Apogee (km)	Perigee (km)	Period (min)	Satellite weight (kg)	Payload weight (kg)	Number of exp.	From countries	Launch date
ESRO-II (Iris)	solar X-rays, and particles; galactic cosmic rays	97.2	1094	323	98.9	75	21	7	F, GB, NL	68–05–17
ESRO-I (Aurorae)	polar ionospheric and auroral phenomena	93.8	1534	253	103	86	21	8	DK, GB, N, S	68–10–03
HEOS-I	interplanetary magnetic fields, solar wind, energetic solar particles	28.3	223428	424	6749	105	27	7	B, D, F, GB, I	68–12–05
ESRO-I (Boreas)	polar ionospheric and auroral phe nomena	85.1	393	306	91	86	21	8	DK, GB, N, S	69–10–01
HEOS-A2	interplanetary magnetic fields, solar wind, solar particles	90	238119	359	6840	117	30.5	7	D, DK, F, GB, I and ESTeC	72–01–31
TD1A	stellar UV spectrophotometry, solar and non-solar X and γ-rays, cosmic rays	97.6	545	533	97	471	144	7	B, D, F, GB, I, NL	72–03–12

III. Some of the Main Results

It is impossible to give in a brief review as the present one even a highly cursory summary of the main results, even if we restrict ourselves in this paper to the main results in the field of astronomical research. We shall only mention some subjects, and refer further to the excellent *Annual Reports of ESRO to COSPAR.*

Solar wind: The two HEOS satellites (Fig. 1) which move partly out into interplanetary space have rendered valuable information on the solar wind: its energy spectra in many channels, its bulk velocity and scatter diagrams of the velocity, its aberration with regard to the radial direction to the Sun, and the magnetic field vector. Particularly interesting are observations of the passages of interplanetary shock fronts related to solar particle disturbances across the satellite, where usually the scalar value of the field may change considerably, but not so its direction (Fig. 2).

High-energy emissions of solar flares have been studied intensively. The ESRO II satellite contained instruments for the detection of soft X-rays in several channels

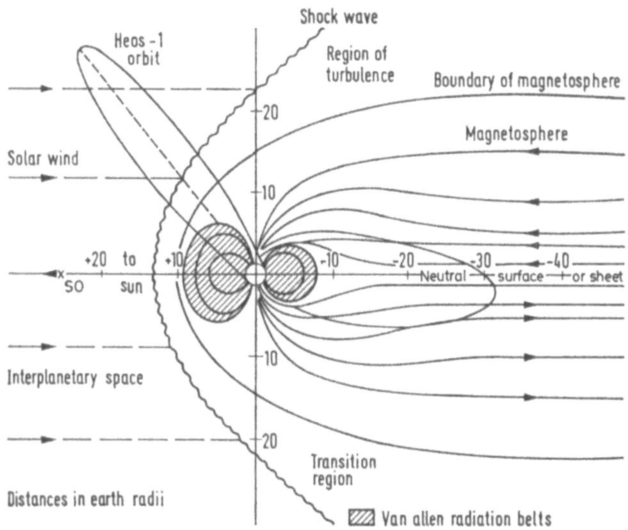

Fig. 1. The orbit of HEOS-I with regard to a magnetospheric model

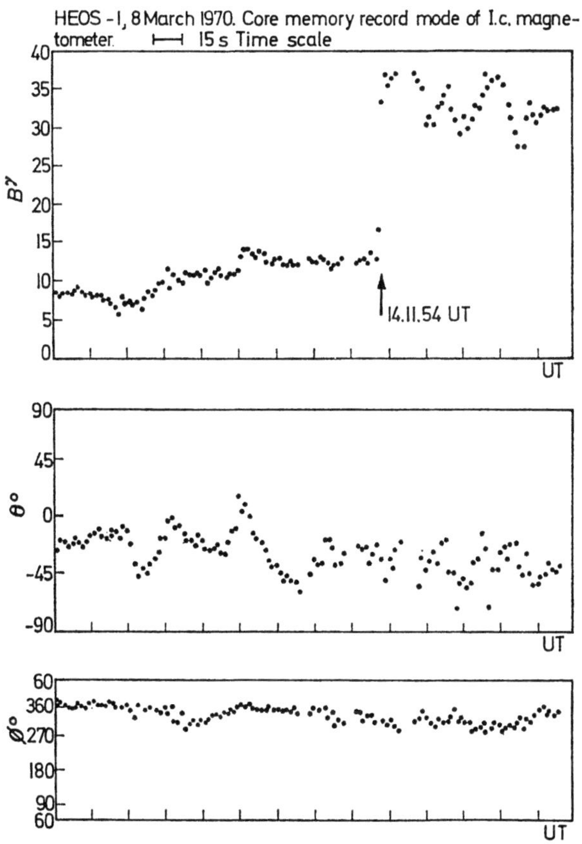

Fig. 2. Details of the arrival of an interplanetary shock front, observed with HEOS-I

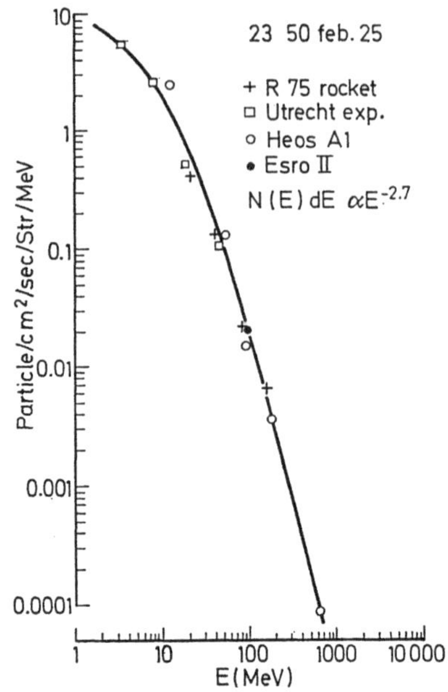

Fig. 3. Particle flux spectra of the event of 25 February 1969, observed simultaneously in interplanetary space and in the polar ionosphere

between a few and 60 Å; the TD1A satellite has a spectrometer for the hard X-ray range. The HEOS and ESRO II satellites enabled the study of the arrival of solar-particle clouds in interplanetary space and in the polar magnetosphere. Particularly successfully was the simultaneous study of solar energetic particles in interplanetary space and in the magnetosphere by means of observations in the HEOS I, ESRO II, and BOREAS satellites, and with rockets launched into the polar ionosphere. The results were discussed in a colloquium, the proceedings of which were published under the title "Intercorrelated satellite observations related to solar events" (V. MANNO and D. E. PAGE, editors; published by D. REIDEL, Dordrecht, 1970). The colloquium concentrated on the study of *one* series of solar events, occurring around 25 February 1969. Since the approximately hundred participants represented virtually all important space research institutes from over the world involved in problems of solar-terrestrial physics, the results of certain U.S.A. and U.S.S.R. satellites could also be incorporated in the discussion. It was, for instance, interesting to see that the particle spectra measured in interplanetary space (HEOS), in the exosphere (ESRO II) and in the polar ionosphere (rockets) were virtually identical, shedding new light upon the entry of energetic particles in the polar magnetosphere (Fig. 3).

Stellar ultra-violet radiation: A particular highlight was the recent launching (12 March 1972) of the large TD1A satellite which contains among various other instruments the Liège-Culham telescope for recording broad-band stellar spectral intensities between 1100–3000 Å and the Utrecht spectrophotometer for high resolution (1.8 Å) spectra of bright stars between 2100–2900 Å. Already in the few

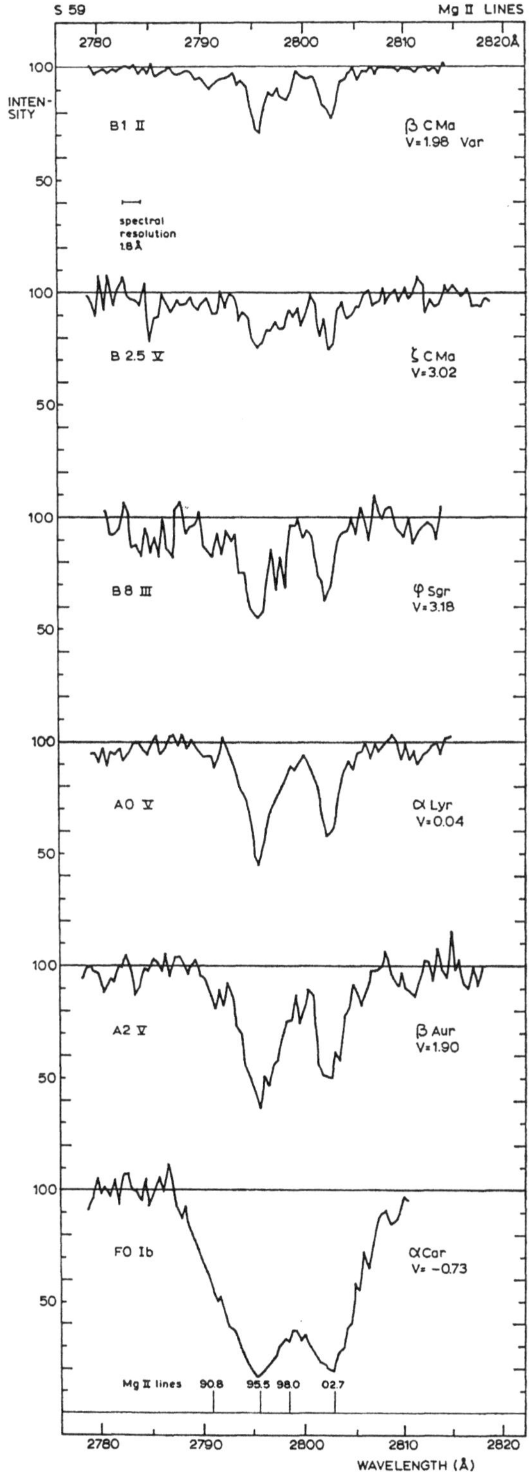

Fig. 4. Profiles of the Mg II lines at 2800 Å in stars of different spectral classification

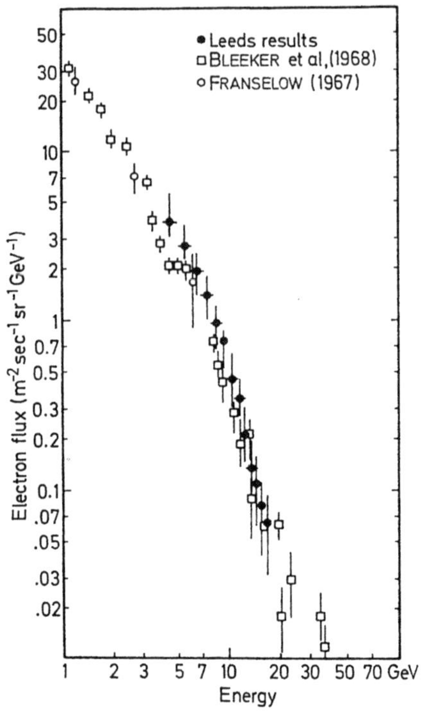

Fig. 5. Energy spectra of electrons in the cosmic radiation, as determined from the observation of ESRO I satellite, and compared with other sources

months immediately following the launching, several interesting results could be published e.g. on the energy-wavelength curve of early-type stars, the behaviour of the UV MgII doublet stars of all early-type spectral classes, (Fig. 4), the occurrence of interstellar lines of different degrees of ionisation in the stellar spectra ... These observations, in which about fourty groups of guest observers from elsewhere in Europe and abroad have shown interest, will certainly prove to give a strong impetus to European stellar and interstellar research.

Cosmic radiation was studied with instruments aboard the ESRO II, HEOS and TD1A satellites. The observations have allowed among other things a sharper definition of the energy spectrum of cosmic-ray particles (Fig. 5).

IV. The Future

Some data on satellites in preparation (summer 1972) are shown in Table 2.

A unique case is the COS-B satellite, in which *one* large instrument for the detection of extraterrestrial gamma rays will be constructed by groups from *five* different institutes, in close mutual cooperation. We should also mention the *geostationary satellite GEOS*, which has been designated as the reference satellite for the International Magnetospheric Survey (I.M.S.), 1966–1968.

Table 2. Satellites in preparation for future launch (this table was established in summer 1972)

	Main scientific program	Inclination (°)	Apogee (km)	Perigee (km)	Period (min)	Satellite weight (kg)	Payload weight (kg)	Number of exp.	From countries	Launch date
ꞌO-IV	ionosphere, solar particles in the polar magnetosphere	90	1100	280	97	113	32	6	D, GB, NL, S and ESTeC	Nov. '72
S-B	gamma rays	25	100000	350	2230	280	115	1	D, F, I, NL and ESTeC	1974
OS	particle fluxes, electric and magnetic fields in magnetosphere (geostationary satellites)	0			day	180	31	9	CH, D, DK, F, GB, I and ESTeC	1976

Three enterprises are actually under consideration in whichcooperation with NASA is foreseen. The first is the *International Ultraviolet Explorer* (formerly called SAS-D), actually a cooperation between NASA, the U.K., and ESRO. The latter's contribution will be the establishment of a ground station, available to the European astronomical community during 1/6 of the observing time. At that ground station astrophysicists may obtain spectra and stars to be selected by them, between 1000 and 3000 Å with a spectral resolution of 0.1 Å for stars brighter than $m=6$. Weaker stars will also be observable, with a degraded spectral resolution. The time necessary for obtaining a full high-resolution spectra of stars of $m \approx 6$ will be of the order of half an hour. The expected life time of the I.U.E. will be a few years, during which period an extremely powerful, versatile and unique instrument may be available for furnishing an enormous amount of ultra-violet spectral data to the astrophysicists. Launching of this instrument is foreseen for 1976.

The other enterprise is the socalled *heliocentric mother-daughter combination* (H-MD), consisting of an interplanetary probe, and two magnetospheric satellites in eccentric orbits, separated by variable distances, never exceeding 5000 km, for studying particle and plasma physics in the magnetosphere. The heliocentric vehicle will be launched at the same time as the other two. The scientific program of the three satellites should be strongly coordinated. One of the magnetospheric satellites should be constructed by ESRO, the two others by NASA. The launching is foreseen for appr. 1976–1978.

Discussions were also held with NASA for a possible cooperation of ESRO in the *Venus orbiter project, 1978*. In the field of X-ray astrophysics studies are now well under way on the mission definition and feasibility of an X-ray satellite in an excentric orbit using lunar occultations for position determinations of X-ray sources (HELOS).

Looking into a still further future (≈ 1980), brief reference is made to the socalled shuttle-period (post-Apollo-Period) in which the U.S. space shuttle will become available. This facility will offer the possibility to bring powerful instruments for a brief period (days to weeks) in orbit, if possible to return observations back to Earth, yielding practically similar possibilities as those offered now by conventional observatories. A few study groups for studying specific European contributions to the post-Apollo period are already functioning, and more will be established.

Altogether we may be allowed to conclude, that notwithstanding the present-days difficult economic situation in many countries and the recent reduction in the scientific part of the ESRO program, ESRO will continue to offer a number of fascinating possibilities to European scientists. How these possibilities will be realized will depend for a great deal on efforts and inventivity.

Discussion

OGELMAN, H.:
How dependend is ESRO on NASA for launching facilities and vehicles?

DE JAGER, C.:
Up to now all ESRO satellites, were launched by NASA rockets.

GRATTON, L.:
1. Can you tell something about ESRO projects for X-ray astronomy?

2. To your knowledge are there existing European projects (in ESRO or outside ESRO) considering the use of focussing systems in X-ray observations (solar and non-solar)?

DE JAGER, C.:
1. An interesting project, now in consideration, is HELOS, an excentric orbit satellite, that should continually point at the moon, and that should use the occultations of X-ray sources behind the moon for precise position determinations.

2. Although high collectors are considered (e.g. in the HELOS project), no focussing systems are up to now considered in the ESRO satellite projects.

Joint Organization for Solar Observations (JOSO)

By K. O. Kiepenheuer
Fraunhofer Institut Freiburg i. Br., Germany

With 1 Figure

Summary

The final goal of JOSO is to find an optimal solar observing site and then to install there a cooperative european observatory. A systematic site testing program is under way since 1969, with the cooperation of most of the european countries. The basic atmospheric concept of this program is the recently confirmed existence of a highly homogeneous airmass above the oceans. The procedures used are optical and, to a large extent, those used by meteorology and atmospheric physics.

The study of a number of coastal and island sites has shown that it is difficult to get direct optical access to this airmass, mostly because of local convection up to a few hundred meters altitude.

A recent investigation of the airmass above the Canary islands (Tenerife and La Palma) by using a specially equipped aircraft for recording in situ the degree of homogeneity of the atmosphere above the islands, has brought a number of significant results about the interaction of the maritime air with the topography of the islands. It was found that the top of the Pico del Teide, an extinct volcano of 3718 m altitude, seems to be almost continuously in direct touch with an undisturbed airmass and thereby probably the only site in Europe (as Hawaii for the US) which will guarantee long periods of undisturbed day-time seeing.

In contrast to ESO which tries to provide optimal observing conditions for the southern night sky and to ESRO which offers extraterrestrial research facilities to the european community, JOSO – in a cooperative effort of european solar astronomers – is searching systematically for better solar observing conditions on the ground, eventually aiming at the creation of a large cooperative observatory.

At the time being JOSO represents a kind of alliance of almost all active european solar astronomers, with the common goal of improving their observing conditions and facilities on the ground. In the effort to find an optimal observing site, Germany, France, Italy, the Netherlands, Norway, Portugal, Spain and Sweden have contributed considerably, to a lesser extent also Greece, England, Israel and Switzerland. Within JOSO the member countries are represented in a provisional board which has a chairman (K. O. Kiepenheuer) and a secretary (C. de Jager). There are various working groups, dealing with problems of testing technique, meteorology, finances and administration. The organisation which meets once a year exists since 1969.

It has been tried to reach for JOSO a kind of european legalisation. The most promising opportunity would have been to connect it with the ESO organisation.

But it soon became clear that this possibility had to be discarded for practical, financial and formal reasons. The aim of this legalisation is of course to find a permanent governmental assistance of all member-countries. It is now attempted to reach this goal stepwise, and first by a convention between the various national sponsoring foundations. Rather promising contacts have been accomplished already between the Deutsche Forschungsgemeinschaft (Bonn) and the Centre National de la Recherche Scientifique (Paris).

Motivations

The sun is the only star which offers to us a stellar atmosphere in which physical phenomena can be isolated and can be investigated individually. Most of them are of magnetohydrodynamic nature and occur within volumes of $\lesssim 1$ arcsec dimension. These structures are mostly characterized by a scale height or scale length of the order of 100 km, corresponding to an angle of ~ 0.15 arcsec. Quite in contrast to this requirement, the very few best solar spectra obtained on the ground show an angular resolution of only ~ 0.7 arcsec. On the average – under "good" conditions a resolution of a few arcsec is obtained! Time sequences of spectra of better than 1 arcsec resolution are extremely rare. The most interesting parts of the solar atmosphere therefore appear blurred to us. The observations do not allow to prove or disprove certain theoretical models of granules, umbral and penumbral structures in sunspots, of various chromospheric structures etc.

The reason for this deplorable situation is not only that our day time atmosphere is usually more disturbed by convection than at night, but also that the majority of solar observatories has been placed on sites which were selected rather according to logistic or diffuse meteorologic criteria than to modern knowledge of atmospheric physics.

In principle it is true, of course, that these seeing problems can be overcome by placing the telescope in an orbit outside the earth's atmosphere or by observing in the upper stratosphere from unmanned balloon borne equipment. A number of high resolution balloon borne telescopes have indeed been built and flown. They have brought photographs of the solar disk with a resolution of ca 0.3 arcsec, but no spectra as yet of this resolution (exposure time about 1000 times longer than for photospheric photographs!). Projects aiming at this difficult goal are under way. But it must be emphasized that this is still an extremely expensive tool of research. The cost of one flight with ca 8 observing hours might come up to the price of a medium sized modern solar observatory on the ground. In space the situation is still more difficult. The advance into the 0.1 to 0.2 arcsec regime in the visible and in the EUV will probably be reserved to Post Apollo technology, to a free flying solar module, maintained and operated through a space shuttle. In any case we will have to wait very probably until the end of this decade. The cost of such a project will exceed the cost of modern ground telescopes by several orders of magnitude.

It is obvious that under such circumstances ground based observing conditions should be exploited systematically to their very optimum. We have learned in the last years that above the oceans, far enough away from a continent, an airmass must exist which shows very little convection and therefore possesses a high degree of optical homogeneity as compared to the airmass over a continent. This expectation has been fully confirmed recently by aircraft studies above the Atlantic ocean near the Canary islands, about 500 km away from the african coast. By using a special sensor, air temperatures could be recorded with a time resolution of 0.002 sec and an accuracy of 0.01°. Thus Schlieren down to a diameter of ~ 10 cm could be recorded during flight (flying speed 150 km/h). The average temperature fluctuations in altitudes up to 5000 m amounted to $\lesssim 0.01 \ldots 0.02°$ under stable weather conditions. This signifies that in such an airmass even a telescope of a few meters diameter would yield its theoretical resolution.[1]

But how can one get a direct optical access to this airmass? To settle this question is and will be the main task of the many JOSO site-testing campaigns which are now underway since 1969.

Results of Testing Campaign

There is no need to report here in detail about the inspections and the many meteorological and optical tests which have been conducted in the last years in the Mediterranean and along the atlantic coast. They all had the purpose to reach the undisturbed atmosphere over the sea. In doing so, we found out that even on coastal sites with a steady sea breeze (bringing undisturbed air to the site) image deterioration was observed due to the occurrence of an anti-sea breeze above some 500 m altitude, produced by the sun's heating of the coastal area. At a few sites of this type, especially in southern Portugal, studies are continued. Even on islands this disturbing effect can be observed. It disappears only on very small and flat islands, too small indeed to house, to operate a larger observatory. Such sites so near to the sea show, of course, a rather high level of scattered light. The strong corrosion of the instruments by the salty air in addition will be a disturbing factor.

The choice of such maritime low level sites is furthermore narrowed down by the condition that there should be no mountains on the upwind side over several hundred kilometers. Otherwise the orographic disturbances would influence the image quality.

For all these reasons we came to the conclusion that it would be best to investigate thoroughly a more transparent and characteristic case within the large and stable airmass of the subtropical anticyclone of the Azores above the Atlantic Ocean, far away from the african west coast. Our main objective was the interaction of this undisturbed airmass with the two mountainous islands of the Canary group, Tenerife and La Palma. The meteorological situation is characterized by the occurrence of a descending air motion (subsidence) in the troposphere below 5000 to 6000 m which causes the formation of a quasi persistent temperature inversion at about 1000 m altitude, mostly connected with a cloud layer. This inversion layer acts as a barrier to the convection produced below it.

---

[1] See e.g. U. Grossmann-Doerth, Solar Physics **9**, 210 (1969).

The survey of the airmass has been conducted by means of a slow flying aircraft[2] Dornier 27 (Sky Servant). On a total tract of about 5000 km, at altitudes between sea level and 5000 m, the following results have been obtained, which may have a significant effect on the future of the JOSO project:

1) As already mentioned, the high degree of homogeneity of the airmass above the Atlantic and above the inversion layer could be confirmed up to 5000 m altitude (ceiling of aircraft). Temperature fluctuations did not exceed 0.01°, which is the noise level of the equipment.

2) On the island of Tenerife above the mountain ridge extending from NE to SW (average altitude ~2000 m, total length ~50 km) usually a turbulent layer of 500 to 1000 m thickness was observed during NW wind (surface wind speed on the ridge >5 m/sec), especially on the lee side of the ridge. Consequently the day time image quality at the Spanish solar observatory Izaña, which is placed at the crest was sensibly reduced. An example is given in Fig. 1.

During the rare periods with a wind speed <2 m/sec this disturbance was absent, and the image quality was then very good. For these optical tests at Izaña a vacuum reflector of 40 cm aperture (Fraunhofer Institut, Freiburg, Germany) was used.

Similar results have been obtained at the island of La Palma above the volcanic crater of La Caldera. Also an optical test on the Muchachos (2426 m) was conducted with a telescope of 20 cm aperture. (Mixed observing group of Göttingen, Oslo and Stockholm Observatories).

3) Conditions around the extinct Volcano Pico del Teide (3718 m), also on the island of Tenerife, were found to be different. The Teide is the highest elevation within the Azores' anticyclonic system. Its top extends about 1400 m above the mountain ridge of the island. From its simple round and conical shape it may be expected that the streaming pattern around it might also be simple.

A number of scanning flights, some of them very close to the peak (minimum distance ~30 m) have shown indeed, that during long periods the top sticks practically out of the perturbation produced by the mountain, especially on the upwind side (mostly W to NW), while the lee perturbation mostly on the S or SE side does not reach high enough to get into the line of sight to the sun. An example of a record is given in Fig. 1. Only for wind speeds >20 m/sec the lee perturbations are significant. These preliminary results – no ground data, neither optical nor meteorological are available yet – make it probable that the Pico del Teide may be the only site for Europe which is in direct contact with an airmass of extreme homogeneity during a considerable fraction of time. A similar situation will only be found again at Hawaii. Another advantage of the Teide site is, of course, its enormous amount of sunshine and its low level of scattered light. The investigation of this site will be continued thoroughly.

At this occasion still a word of caution: Our aircraft results end at 5000 m altitude. We have good reasons to assume – at least for a site on the Atlantic ocean far away from the continent – that the good conditions found at 5000 m will

[2] Project of the Fraunhofer Institut, Freiburg, in cooperation with the Deutsche Forschungs- und Versuchsanstalt für Luft- und Raumfahrt, Oberpfaffenhofen, with financial aid of the Deutsche Forschungsgemeinschaft.

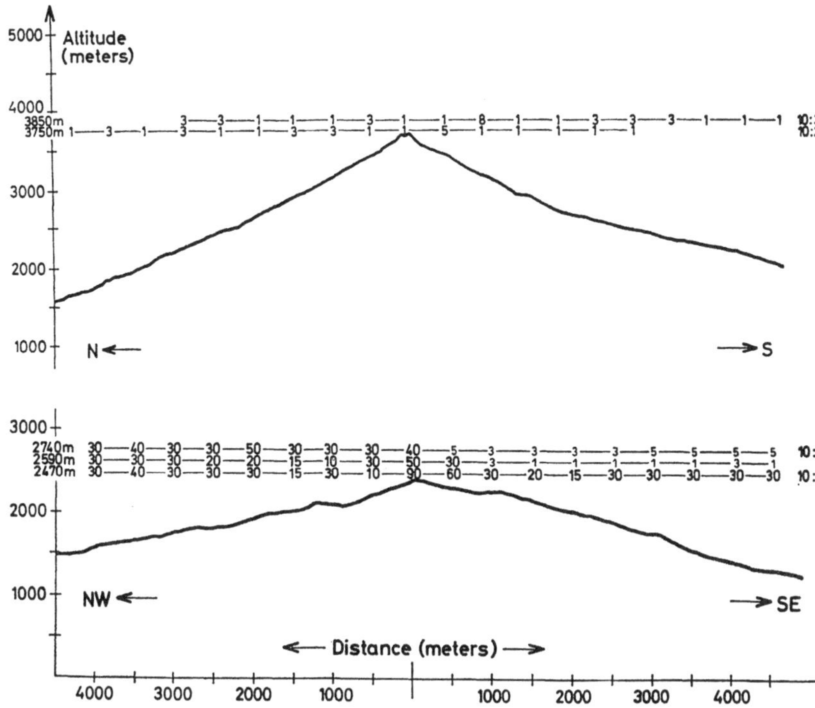

Fig. 1. Typical records of air temperature fluctuations obtained from an aircraft on July 15, 1972, in units of 0.01° C at a flying speed of 50 m/sec. Altitude of aircraft is marked on the left, flying time on the right. *Above:* Profile of Pico del Teide (3718 m), west wind 8 to 10 m/sec. *Below:* Profile of Montagna de Izaña (2400 m with Izaña observatory on top), NW wind 6 to 8 m/sec.

While above the Teide the airmass turns out to be extremely homogeneous (ΔT mostly $\lesssim 0.03°$) down to about 30 m above the peak, the records above Izaña evidence clearly a disturbed layer (ΔT 0.3 to 0.9°) of a few hundred meters thickness

continue upwards. Above a continental site, in particular over a mountainous topography, this quite often will not be so. Nevertheless we will be aiming to get in situ results also from above 5000 m up to the tropopause also for the Canary sites. On the whole, there can be no doubt that also these higher parts of our atmosphere will interfere occasionally, although not too often, in the case of an ocean site.

Outlook

The final goal of JOSO is to create observing conditions which permit a resolution comparable to the solar scale-height during a reasonable fraction of time, to obtain occasionally spectra ($\lambda/\Delta\lambda > 400000$) with an angular resolution of this order. This implies of course the occurrence of even longer periods with a resolution better ~ 0.4 arcsec. It should not be overlooked that the possibility of obtaining frequently spectra with 0.4 arcsec resolution would already bring an enormous increase of information and a great progress in interpreting the solar atmosphere.

It is difficult at the moment – in the midst of the campaign – to set resolution limits for the final choice of a site. The testing campaign will certainly go on for at least two more years. Priority will be given to the Teide, to a mountain site on the island of La Palma and some coastal sites in southern Portugal (Barreta/Faro). Simultaneously with this continuous integration of seeing experiences the diverging criteria for a final JOSO site will hopefully converge. The chances to find a site which is decisively better than existing observatory sites have improved considerably during the campaign.

If the site is found and the choice is agreed upon, the first action will be to transfer existing medium sized telescopes, now at poor sites, to the new site. Especially Germany and France dispose of instruments available for this purpose. This relatively economic procedure does not only take account of the somewhat difficult financial situation in Europe, it also enables us to test the site thoroughly before building and installing a large solar instrument of $\gtrsim$150 cm aperture.

Discussion

WEHINGER, P.:
Have you considered building a solar observatory on a mountain lake like Harold Zirin Big Bear Lake Observatory?

KIEPENHEUER, K. O.:
We have been looking into this possibility. But in order to eliminate completely the effect of the surrounding mountains this lake should have a size of more than 100 km.

BORGMAN, J.:
The selection of La Silla (ESO) was also partly based on a study of microturbulence? Siedentopf (Tübingen) had several towers on the mountain to measure the height dependence of micro-turbulence by fast thermo-elements.

KIEPENHEUER, K. O.:
Measurements of microturbulence within the first 20 to 40 m above ground level give unfortunatly only a very incomplete picture of the disturbances occuring over the site, while recordings from an aircraft will cover all disturbances existing along the line of sight.

McCARTHY, M. F.:
Can you tell us something about the winds near mount Teide (3700 m) region? Are measurements being made of wind velocities on Tenerife near this site?

KIEPENHEUER, K. O.:
At the moment we have only the daily radio balloon data obtained at Santa Cruz de Tenerife. We are joing to install at the Pico now a recording anemometer.

MEZGER, P. G.:
You have found or you expect to find from aircraft tests on the Canary Islands what sounds like the world's best seeing conditions (0.15). Is this also true for night observations? If yes, would this the be a site better than the ESO site in Chile?

KIEPENHEUER, K. O.:
The excellent day-time observing conditions to be expected on the Pico Teide would probably also exist during the night. Whether they are better than at the ESO site may be found out only after thorough optical tests. The value of 0.15 which you mention is of course very optimistic.

SMYTH, M. J.:
Have you also investigated la Palma in the Canary Islands, which has a ridge over 2000 m but no central peak?

KIEPENHEUER, K. O.:
Yes, we have also investigated La Palma, especially the Pico Muchacho (2400 m). As aircraft investigations show, this site seems to be slightly better than the Izaña sites.

The Projects of the I.N.A.G. (France)

By J. Delhaye
INAG-Observatoire de Meudon, Meudon, France

Summary

The activities of the National Institute for Astronomy and Geophysics (Institut National d'Astronomie et de Géophysique, INAG) are briefly described and its principal current projects presented: the creation of a Center for studies and research in geodynamics and astronomy, the construction of the Bernard Lyot telescope (2 m) and of a Schmidt telescope (1.5 m mirror, 0.9 m plate), and plans for a 3.6 m telescope, are now in an advanced stage of preparation. Other, longer-term projects are mentioned, in particular an interferometer to operate at mm wavelengths and an infrared collector.

The National Institute for Astronomy and Geophysics (INAG) was created in the autumn of 1967 and began operating at the beginning of 1968.

The aims of INAG are the development and coordination at a national level of research in astronomy and geophysics which are being carried out in the various observatories and laboratories attached either to the universities or to the National Center for Scientific Research (Centre National de la Recherche Scientifique, C.N.R.S.).

INAG operates a certain number of research programs, is responsible for national projects (as well as those which exceed the capacity of individual laboratories), can create new research units should the necessity arise, and supervises research contracts (in particular, in the field of space research).

The activities of INAG are limited to coordination and support and the Institute in no way usurps the authority of the individual laboratories, nor their autonomy in matters of personnel, administration of the laboratories, and choice of research projects. INAG itself carries out only technical research.

We review below the main projects of the Institute. Those which are already under way will be briefly described, others will simply be mentioned.

The "C.E.R.G.A." (Centre d'études et de recherches géodynamiques et astronomiques)

Positional astronomy is often carried out, in France as in most other countries, in outdated observatories, built in close proximity to university towns, into which they have long since been absorbed. The aim in creating a center for studies and research in geodynamics and astronomy is to group together the observational instruments and research workers at a site which is sufficiently protected, but not as isolated as those often chosen for astrophysical observatories.

Investigations for a suitable site began in 1965, and eight possible sites were tested by making observations with a Danjon astrolabe, generally over a period of about one year. The results of these observations were compared with those obtained at the Observatories of Besançon and Paris where astrolabes have been operational for many years. The best site was found to be that on the plateau of Caussols, which is 10 km long and a few km wide, and is situated 10 km north of Grasse, in the Alps, at an altitude of approximately 1300 m. At this site, there are about 160 completely clear nights each year, and 70 or 80 nights which are partially clear and can be used for astrometry. Observations made with a polar telescope (the technique of observing the trail of the pole star) confirmed the results obtained with the astrolabe, the size of the displacement of the image being of the same order of magnitude as the random errors in the observations of stars with the astrolabe.

By 1975, CERGA will be equipped with: an astrolabe (already installed), a laser for lunar telemetry (enabling distances to be measured to an accuracy of less than 1 m), and a Schmidt telescope with a 150 cm mirror and a 90 cm plate. It is also hoped that a photoelectric astrolabe will be built by this time, operating at both 25° and 40° zenith distances and enabling a catalogue of absolute declinations to be compiled. Finally, it is intended that a satellite tracking station will be built and that a natural grotto will be fitted out to take horizontal pendula for the study of terrestrial tides.

One of the basic ideas of the project is that it is essential to bring together at the Center the observers, the engineers who are interested in the development of the instrumentation, and the research workers. The activities of the latter include all aspects of the relative motions of astronomical bodies, including the Earth.

The Bernard Lyot Telescope

This telescope, which will be 2 m in diameter, is intended to attain the highest spatial resolution possible from the ground and, with this objective, will be placed at the Pic du Midi. There, the excellent atmospheric conditions in view should enable us to attain an angular resolution of better than $0''.2$. It will be used to observe the planets, *point* objects at the limit of detection, close visual binaries (by means of an automatic image selection technique) and to make detailed studies of extended objects. The primary mirror is made of Cer-vit. The aperture will be $f/5.0$. In addition to the usual foci, the Bernard Lyot telescope will possess a Strand focus. The instrument is at present under construction.

The 1.5 m Schmidt Telescope

This telescope has already been mentioned in connection with the equipment of CERGA where it is intended, at least provisionally, to be installed. Its characteristics are as follows: a mirror of Cer-vit of 152 cm diameter (150 cm of which can be used); a 94 cm silica plate (of which 90 cm is usable); a focal length

$f = 3.155$ m (giving an aperture ratio of 3.5); and a field of 6° × 6°. The mirror has already been polished and the plate is in the process of being polished. This telescope will be used mainly for astrophysical observations, but it is hoped that, in view of the careful considerations given to the conception of the instrument, it will be possible to use it as a wide-field astrograph.

The 3.60 m Telescope

The construction of this telescope is the most important commitment of INAG in the field of astronomy. The search for a site began in 1965 (before INAG was created) and lasted six years. Sites were studied in continental France, Corsica, southern Spain, the Canary Islands, Mexico and Hawaii. Finally, at the beginning of 1972, Mauna Kea in Hawaii was selected (although no official decision has yet been taken).

The ambition of those in charge of the project is to create an instrument of high optical quality which will enable very faint objects to be studied with an optimum signal-to-noise ratio and also permit wide field observations to be made with a high limiting magnitude. Furthermore, the properties of the chosen site will result from a high priority being given to infra-red observations, as the 3.60 m telescope will be the largest located in a site which is well suited to this type of work.

An additional objective is the design of a minimum light loss coudé focus. As a consequence, the coudé focus will be of small aperture and will employ plane secondary mirrors having a small diameter, interchangeable, and can be coated to obtain a reflectivity of 95% in each wavelength range.

Finally, it is desirable that the telescope should be adaptable to satisfy needs which might arise in the future but are difficult to foresee. This flexibility is assured by the possibility of changing the upper part of the telescope.

At the present state of the project, the anticipated characteristics of the telescope are as follows:

Primary Mirror; parabolic, made of Cer-vit; aperture ratio 3.8 ($f = 13.68$ m); without a corrector, the field will be 2′.5 (10 mm) and the coma spread less than 1″; the field will be able to be increased to 15′ (60 mm) with a 3-lens corrector and to 1° (240 mm) with 4 lenses.

Visual Cassegrain Focus; aperture 7.6 ($f = 27.36$ m) giving a field of 10′ (80 mm) which could be increased to 40′ (320 mm) with a corrector.

Infra-red Cassegrain Focus; the secondary mirror will be quite small (20 cm) in order that it may be vibrated at quite high frequency; in addition, its focal length will be such that the diffraction pattern at 10 microns will have approximately the same diameter as the receiver (this condition will be attained by an aperture of 70 ($f = 252$ m, resulting in a field of 10′ or 800 mm).

Coudé Focus; the low latitude of the site and certain structural considerations have given rise to the conception of a coudé focus using 5 or 6 mirrors (small, if possible). One is, therefore, led to adopting a system (similar to that at the Observatory of Victoria) of a secondary mirror with an aperture of $f/90$ and made up of 3 coated mirrors mounted on a turret: mirrors 3, 4, 5 and possible 6 will also be composed of 3 coated mirrors on turrets. The mean size of these mirrors is 25 cm.

A field lens and a 3-lens objective having good transmission in the ultra-violet and near infra-red will produce a beam of aperture *f*/25; the unvignetted field will be 2′.

Other Projects

The realization of the above projects is already under way or shortly will be, but INAG has others under consideration as well. They include an interferometer to operate at mm wavelengths and a large infrared collector, but it would be premature to try to specify their characteristics. Similar but smaller instruments are already operating (a millimetric interferometer at Bordeaux and an infrared collector with a mosaic mirror at Meudon) and on the larger ones will be designed on the basis of information obtained from these instruments. The sites have not yet been chosen.

We should also mention that studies of receivers and methods of data analysis, which are of more general interest, are also under way.

Discussion

ULRICH, MARIE-HÉLÈNE:
Although these are not projects for the near future, could you already say what would be the shortest wavelength at which the millimetric interferometer and the I.R. collector could be used?

DELHAYE, J.:
As regards the millimetric interferometer, it seems highly desirable that it should be able to function correctly down to the 3 mm atmospheric window. At still shorter wavelengths, the performance will undoubtedly depend greatly upon the quality of the chosen site.

As for the infra-red collector, one proposal is that the instrument should be able to be used right into the visible, at least for certain spectroscopic work.

SMYTH, M. J.:
What aperture is envisaged for the large infrared flux collector?

DELHAYE, J.:
Preliminary studies carried out under the leadership of Dr. P. CONNES, consider an aperture of about 20 m, but the cost of building such an instrument would be high. An instrument of a similar type but with a smaller diameter of 10 or 12 m would still be of considerable interest.

FOKKER, A. D.:
Will the interesting new projects envisaged by INAG have any influence on the extent to which France will continue to contribute to ESO?

DELHAYE, J.:
It is clear that INAG must take into account French international commitments. Furthermore, the French astronomical community attaches the greatest importance to the normal development of the activities of ESO.

The Status of the Max-Planck-Institut für Astronomie, Heidelberg

By WALTER FRICKE
Heidelberg, Germany

In reporting on the foundation of the *Max-Planck-Institut für Astronomie* at Heidelberg, reference is made to previous reports given by ELSÄSSER (1971, 1972). The establishment of the new Institute has been based mainly on the following considerations: During the past twenty years the astronomers in the Federal Republic of Germany have agreed, first, that at the existing observatories in West Germany no installment of large telescopes can be recommended due to poor observing conditions, and, second, that, besides the participation of the F.R.G. in the European Southern Observatory, some powerful telescopes should be made available to German astronomers and operated under their own responsibility in some foreign countries with favorable observing conditions.

The *Deutsche Forschungsgemeinschaft*, in its "Denkschrift zur Lage der Astronomie" (1962), and the *Wissenschaftsrat* of the F.R.G. have supported these ideas. On this basis the *Max-Planck-Gesellschaft zur Förderung der Wissenschaften* founded the *Max-Planck-Institut für Astronomie* at Heidelberg in 1969 and appointed Professor ELSÄSSER as its first director. He is advised by a Scientific Council with Professor STRÖMGREN as its Chairman. At this conference, a few other Council members besides myself are present, and I may mention Professor BORGMAN in particular, since he can answer any questions on details of the telescopes which are being designed at present.

The Institute will have its headquarters in Heidelberg, where buildings are under construction for offices and laboratories. The first buildings will, we hope, be completed in 1974. An observing site has been found in Southern Spain at an altitude of about 2200 m near Almeria, where three medium size telescopes will be installed. This past summer an official agreement was been signed by the government of Spain and the Max-Planck-Gesellschaft for the lease of the land for the station. Right now, Professor ELSÄSSER and his colleagues are in Spain making arrangements for the construction of buildings and domes on the mountain. The station will be equipped with three telescopes: (1) one photometric telescope of 129 cm aperture which has already been completed by Zeiss, Oberkochen; (2) one spectroscopic telescope of 220 cm aperture, and (3) the Schmidt telescope of 80 cm aperture now operated at the Hamburg Observatory. The users of the telescopes will be the members of the permanent staff of the Max-Planck-Institut and of other observatories in the F.R.G., astronomers from Spain, and scientific guests of the Max-Planck-Institut.

Furthermore, a telescope of 350 cm aperture is under design. The Zerodurblank of the mirror is already available, and Zeiss-Oberkochen will build the telescope.

Details of the design are, however, not yet completed. No decision has yet been made concerning the site of this big telescope. Under consideration are sites located in Chile, in South-West Africa near Windhuk, and in Spain.

References

ELSÄSSER, H.: In: Large telescope design. Proc. ESO/Cern-Conf. Genf, Ed. R.M.West p.75, 1971.
ELSÄSSER, H.: In: Auxiliary instrumentation for large telescopes. Proc. ESO/CERN Conf. Genf. Ed. S. Laustsen and A. Reiz p. 83, 1972.

Discussion

VOIGT, H. H.:
As far as I know the MPI will have two telescopes of 2.2-m apparture. Can you give details about these telescopes?

BORGMAN, J.:
Indeed, Prof. VOIGT is right; two 2.2-m telescopes (Cassegrain and Coudé) are in the program; the first ready will be installed at the Calar Alto, the second will go to the southern hemisphere.

CESRA

By A. D. Fokker
Utrecht University Observatory "Sonnenborgh", Utrecht, The Netherlands

CESRA stands for Committee of European Solar Radio Astronomers. The basis for CESRA has been laid at the IAU General Assembly in Brighton, 1970. Its first meeting took place in February 1971 in Utrecht. A second meeting was held in September 1971 in Trieste. In the mean time CESRA has become a section of the Physics in Astronomy Division of the European Physical Society.

CESRA was born out of the need to bring more coherence in the efforts of solar radio astronomers in Europe and out of the conviction that further progress in the study of solar radio phenomena requires observing facilities of such a cost and complexity that they will surpass a single nations possibilities. As a matter of fact, solar radio work is scattered all over Europe, and is pursued mostly by small groups with comparatively limited means. But the object of our research, the Sun, calls for a variety of sophisticated observing facilities. The sun's radio phenomena have interesting and intriguing properties which result from processes that may well prove to be fundamental. Many phenomena occur on an extremely short time scale and their study therefore requires high time resolution. Moreover, there is a great want of two-dimensional representations of brightness distribution. The Culgoora radioheliograph, operating at 80 MHz, has taught us what a wealth of information can be obtained from high resolution heliography.

Further progress in the observational study of the radio sun cannot be expected from single modest observational programs as they have so far been carried out by several small-sized European observatories. In order to achieve in the future something important, a major international project should be undertaken. Otherwise, solar radio astronomy in Europe will, in the course of the next ten years or so, loose all possibility to contribute something valuable to the study of solar radio phenomena. So I arrive at the general purpose of CESRA which I state in 4 points:

1) To stimulate the cooperation between solar radio astronomers in Europe (the cooperation doesn't need to be restricted to Europe).

2) To define and coordinate research programs directed towards solving specific problems of solar radio astrophysics.

3) To recognize and define such types of research as may ultimately lead to a breakthrough in solar (radio)physics, but which require an effort that surpasses a single nations possibilities.

4) On the basis of such an evaluation, to select, to promote and to carry through a major international research project, the scientific importance of which is generally recognized.

The cooperation and exchange of ideas has already been strongly stimulated by the two meetings CESRA has held so far. Later this month (September 1972) CESRA will meet again at Bordeaux.

Specific research programs have so far not yet been formulated explicitly, but a certain consensus about the fundamental issues of solar radio research may well grow out of CESRA's Summer School on "Plasma physics in its application to solar radio astronomy" which is just now being held at the Ile de Ré, in France.

A working group on a Joint Interferometer Project will report to the Bordeaux meeting on the potential scientific merits of a high resolution two-dimensional quick-scan microwave heliograph.

A major international project in solar radio astronomy, whatever this may be, is likely to be competitive with ambitious projects in other, related, disciplines. A number of consultations with scientific and governmental bodies and with industrial firms, will have to be held before any decision on a major solar radio project can be taken.

The most fundamental challenge in solar activity is the flare phenomenon with its related effects. But this challenge is a matter of concern to radio, optical and space researchers alike. And the planning of solar radio work should, accordingly, not be isolated from the question how solar research in general and the study of solar activity in particular, should be developed. CESRA will be glad to help preparing an integrated plan of action for an international effort to tackle the main problems of solar activity. Also for the sake of raising money, a combined solar project is more likely to be succesful than separate optical, radio etc. projects. The activities of JOSO already demonstrate how rewarding it is to combine the available experience and know-how into a common enterprise. CESRA, in its own field, also tries to take the full advantage from having so much skill and knowledge among its members, in an effort to create the right conditions for future fruitful work. Our Sun is such an interesting and vitally important object of research, that it would be a shame not to respond adequately to the challenge with which we are so obviously confronted.

Discussion

BORGMAN, J.:
Is your project beyond the means of the individual countries? Are you competing with similar projects elsewhere in the world?

FOKKER, A.D.:
Any project we might adopt is likely to cost several (or many) millions of German marks. Although perhaps France, or Russia could afford such an expensive enterprise, no sophisticated instrument of the merits envisaged is in the picture as yet. From an ideological point of view, it wouldn't even seem desirable that a single country engages itself in such an enterprise alone.

Our motives are not of a competitive nature; it is just for the sake of creating conditions for a fruitful continuation of our work that we think about possibilities of promising research that should fit well into what is done and what is being planned in the rest of the world.

The "Astronomy and Astrophysics" Division of the European Physical Society

By P. G. MEZGER
Max-Planck-Institut für Radioastronomie, Bonn, Germany

In recent years there has been growing awareness of the importance of promoting contact, at all levels, between astronomers in different European countries and of encouraging European cooperation in the field of astronomy. Many people have consequently felt the need for a society or organisation of European astronomers. The new Astronomy Division of the European Physical Society has been formed in the hope that it can meet this need.

Background Information on the European Physical Society

The E.P.S. was founded in September 1968, its purpose being to promote the advancement of physics in Europe and in neighbouring countries. A successful plenum inaugural meeting of the society took place in Florence in April 1969, but the main scientific work is carried out by specialized divisions, which have considerable autonomy in their organisation. Seven divisions are already established and have all been active in organising and coordinating meetings, summer schools, etc. and in promoting contact between laboratories in different countries. The EPS as a whole has already furthered the interests of European physicists in several ways – notably by rationalising the European journals.

The Astronomy Division

The idea of forming an Astronomy Division was first debated at the EPS inaugural meeting in Florence (where a substantial part of the proceedings was indeed devoted to astronomical topics). Further informal discussions during the IAU General Assembly in Brighton and elsewhere indicated extensive support of the suggestion. Consequently, a group of European astronomers and physicists who were already members of the EPS formally proposed that an Astronomy Division should be formed. This proposal was welcomed by the EPS council, and the new division officially came into being on October 21st 1970, with E. SCHATZMAN (Paris) as Chairman of the Interim Steering Committee and M. J. REES (Cambridge) as Secretary. An election has now been carried out, and the elected board comprises the following twelve scientists:

A. BLAAUW (E.S.O. Hamburg)
M. KUPERUS (Utrecht)
F. PACINI (Frascati)
B. PACZYNSKI (Warsaw)

P. O. LINDBLAD (Stockholm)
G. MARX (Budapest)
P. G. MEZGER (Bonn)
E. R. MUSTEL (Moscow)
M. J. REES (Cambridge)
E. SCHATZMAN (Paris)
D. W. SCIAMA (Oxford)
A. W. WOLFENDALE (Durham)

We believe, for several reasons, that the formation of the EPS came at an opportune moment, and that the requirements and interests of European astronomers can be very effectively covered by an organisation working within the flexible framework of the EPS.

1) It enables astronomers to benefit from the existing facilities and administrative structure of the EPS, and renders unnecessary the setting up of a completely new organisation.

2) The scientific academies of both Eastern and Western countries adhere to the EPS, and it should therefore be relatively easy to promote contact between Eastern and Western European astronomers under EPS auspices.

3) Incorporation with a physical society should help to foster closer links between physics and astronomy. This is increasingly valuable at a time when more and more branches of physics are becoming relevant to astronomy, and the importance of astronomical observations for fundamental physics is being increasingly recognised.

4) Although the Astronomy Division is intended to serve astronomers working in all branches of the subject, the association with physicists is especially valuable for "borderline" subjects (e.g. space physics, cosmic ray research, x-ray astronomy etc.).

Activities and Membership

The enlistment of individual members of the division is at present handicapped by the fact that most astronomers do not belong to their national physical society. Unless their national astronomical societies adhere to the EPS (and we hope some arrangements along these lines can eventually be worked out), individual astronomers have to pay a rather high fee to join the EPS. All the scientific activities of the division, however, are open to non-members. An inaugural meeting of the division took place in Meudon in February 1972 and was attended by over 150 persons from twelve European countries. Although further general meetings of the division will be held periodically, perhaps in conjunction with future European Assemblies of the IAU, it is envisaged that a more useful purpose will be served by sponsoring specialised meetings on topics which concern few workers in any single country, but where Europe, as a whole, possesses considerable strength. This need not, of course, increase the net number of meetings which is already excessive! Other functions of the division will include coordination of summer schools and other educational activities, and the exchange of information between centers of research in different countries. The division has already sponsored an European Conference on Cosmic Ray Modulation, and a Summer School on Planetary Physics. The EPS general meeting being held in Wiesbaden next month (at which the division is organising two sessions of general interest to physicists) will provide the first opportunity for the new board to meet and discuss future activities.

Discussion

BORGMAN, J.:
How do you cut down the audience of a symposium, organized for members, to 40–50?

MEZGER, P. G.:
By not inviting more people and by being tough if people want to invite themselves. In fact, in recent months we have had excellent examples of small and specialized symposia. One was on inter-stellar molecules, in Oberstdorf W-Germany, the other on a workshop on compact HII-regions in Leiden, Netherlands.

JAGER, C.:
It is not correct that IAU symposia and colloquia have unlimited attendance. Each scientific organizing committee of IAU meetings has the right to limit the attendance to an amount set by the S.O.C. We have had a colloquium with an attendance of only 15 persons.

MEZGER, P. G.:
Nothing against IAU symposia, the organization of which is usually excellent and whose attendance is still within reasonable limits. What I and what other people do object to are mammoth meetings like the general assemblies of organizations such as the IAU or URSI.

The Italian Project for a 3.5 m Telescope and the New 1.8 m Telescope of Asiago Observatory

By C. BARBIERI
Astronomical Observatory of Padova-Asiago, Padova, Italy

With 3 Figures

Abstract

This paper reports on the recent developments of the project O.A.N. for a 3.5 m telescope to be located in Italy. After the completion of the drawings of the tube assembly, the mirror cell is under construction at the Officine Galileo (Firenze). The ULE blank has been tested and accepted and the choice for the site has been restricted to two places in South Italy.

The new 182-cm telescope of Padova and Asiago Observatory, at present the largest in Italy, is also briefly described.

I wish here to report briefly on the status of the Italian Project for a 3.5 m telescope to be located in Italy, project known as O.A.N. (Osservatorio Astronomico Nazionale). The project has been developed to the present stage by a Scientific Committee including Drs. G. RIGHINI (Head of Committee), C. BARBIERI, F. BERTOLA, L. GRATTON, M. HACK and L. ROSINO.

The basic configuration, chosen after a discussion of the weather and seeing conditions prevailing in Italy as well as of the astronomical goals of the telescope, is a classic paraboloidal primary mirror with focal ratio $f/4$, with direct access to the prime focus, a Cassegrain focus $f/12$ and a Coudé focus (RIGHINI and FARRELL, 1971). The originally planned $f/30$ configuration for this latter focus has been changed in July 1971 to adopt the small mirror system firstly used at the Dominion Astrophysical Observatory (RICHARDSON 1971). The detailed concept of the telescope has been developed by Dilworth, Secord, Meagher and Ass. Consultant Engineering (Toronto, Canada). At present the drawings for the tube assembly have been completed, including the three interchangeable upper ends (a three mirror turrett replacing the original $f/30$ secondary; also the flat mirror mounting in the centre section had to be changed) and the Cassegrain cage. The completion of the pedestal and fork assembly has to wait for the choice of the site of the telescope. Officine Galileo (Firenze, Italia) act as prime contractor for the project and are presently engaged in the construction of the mirror cell.

The U.L.E. blank was ready at Corning Glass plant in Canton (USA) in 1971. The size of the mirror was about 25 mm larger than expected and therefore slight modifications had to be made to the drawings in order to accomodate this extra size and weight. The blank, which appears to be very good, has been officially tested and accepted in August 1971.

Regarding the site of this telescope, after an initial survey at six stations conducted by Dr. O. BENDINELLI, the choice has been restricted to two places in South Italy: Toppo di Castelgrande (200 km E of Naples, 1250 m above sea level,

Fig. 1. The 182 cm telescope assembled at the Officine Sarti (Bologna, Italia)

latitude 40°48′), and Monte San Salvatore in Sicily (100 km SE of Palermo, 1900 m a.s.l., latitude 37°.5). At the first site a Royal Observatory Edinburgh team is also active for the British National Telescope Project. Dr. R. BARBON is at present in charge of the site-test campaign. The Proceedings of the Committee are available as Rapporto OAN No. 1 and No. 2 and can be obtained at request from the Astronomical Observatory of Padova, Italia.

In the meantime, the Observatory of Padova-Asiago has been engaged in the construction of a 182-cm telescope, which will be the largest in Italy until the 3.5 m is in a position to operate. Figs. 1, 2 and 3 show respectively the telescope assembled at the Officine Sarti (Bologna, Italia) and the dome at Cima Ekar (1350 m a.s.l., 4 km E of Asiago Observatory). This telescope has a primary paraboloidal mirror with focal ratio $f/3$. The prime focus will not be used, only two Cassegrain $f/9$ and $f/16$. Provision has been made also for a Coudé focus but no decision has been taken yet regarding this configuration, mainly for financial reasons. The blanks, in Schott Duran 50 glass have been figured by Grubb-Parson & Co. Ltd (Newcastle, England).

All the mechanical and electrical engineering for this telescope has been done by members of the staff of the Observatory of Padova and Asiago.

I think it worthwhile to stress this point in this meeting because our effort has been greatly facilitated by the friendly cooperation of the astronomical community.

In particular we greatly benefited from drawings, discussion and advice from astronomers and engineers of Kitt Peak National Observatory, McDonald Observatory and Vienna Observatory. In this spirit we extend a warm invitation to our european colleagues to apply for observing time at the telescopes of Asiago.

Fig. 2. The dome looking South. The long window of the conference-room is visible

Fig. 3. Living quarters and garages on the North side. The dome is covered with stainless-steel panels

References

RICHARDSON, E. H.: ESO/CERN Conference on Large Telescopes Design, Ed. R.M. West, p. 179, 1971.

RIGHINI, G., FARRELL, J. G.: ESO/CERN Conference on Large Telescopes Design, Ed. R.M. West, p. 65, 1971.

British Plans

By H. A. Brück
Royal Observatory, Edinburgh, Scotland

Summary

The paper discusses British projects for new observational facilities for astronomy which are presently under construction or are being planned by the Science Research Council. It mentions in particular the new telescopes which are being set up in Australia, namely the joint Anglo-Australian 150-inch reflector, and the British 48-inch Schmidt Camera. Mention is made also of plans for a new major "Northern Hemisphere Observatory" for which site-testing is now carried out by members of the Royal Observatory Edinburgh in Mediterranean and Atlantic areas.

Plans for British facilities for observational astronomy are now coordinated by the Science Research Council which was set up by the U.K. Government in 1965. The Council is directly responsible for the support of the two Royal Observatories at Hertsmonceux and Edinburgh, and indirectly, through the grants it disemburses, for most astronomical research in Universities.

The Council's support of national facilities covers all aspects of observational astronomy, experiments in rockets and satellites as well as telescopes and general instrumentation for optical, infrared and radio astronomy. Recently completed instruments in the last two fields include the 5 km radio telescope at Cambridge and a 60-inch infrared flux collector mounted at Izaña on the island of Tenerife. Plans are being prepared for a larger flux collector with an aperture of about 120 inches and for a radio telescope for the Jodrell Bank Observatory with an aperture of 375 feet.

In the field of optical astronomy planning started with a discussion of new facilities for southern hemisphere astronomy. While all existing facilities are situated in South Africa, new ones are to be set up in Australia.

As far as South Africa itself is concerned, the Science Research Council and the South African Council for Scientific and Industrial Research have agreed to set up jointly a new "South African Astronomical Observatory" with its headquarters at the former Royal Observatory at the Cape and an observing station at Sutherland in the Karroo. The first instrument at this new station will be the 40-inch reflector, formerly at the Cape Observatory. Other instruments are to be transferred there in due course, including the 74-inch telescope of the Radcliffe Observatory which may be moved to Sutherland by 1975. The new Observatory which started its existence on January 1 of this year, will be the joint responsibility of the South African and British Research Councils and its facilities will be available to astronomers of both countries.

Major new facilities in the southern hemisphere are being set up in Australia. The largest of these is the 150-inch 'Anglo-Australian Telescope' which is being financed jointly by the two countries. The telescope is under construction at Siding Springs, 350 miles north west of Sydney, where it is expected to come into operation by the beginning of 1975.

On the same site is being set up as a purely British venture a 48-inch Schmidt camera, similar to the Schmidt camera on Mount Palomar. Rapid progress has been made with the construction of this instrument which is expected to be able to start its survey work in 1973. Discussions between ESO and the Science Research Council have ensured that the work of the ESO 40-inch Schmidt camera in Chile and that of the new 48-inch Schmidt in Australia will be properly coordinated and will lead to close cooperation rather than to duplication of effort.

It is expected that the analysis of the new Schmidt photographs will be greatly assisted by the use of the fast automatic measuring machine Galaxy of the Royal Observatory Edinburgh which will be described and discussed by Dr. V. C. REDDISH later during this conference.

In recent years it has at last become possible in Britain to plan not only new facilities for southern hemisphere astronomy, but to consider seriously the establishment of a major new observatory in the northern hemisphere at a site with a sufficiently good climate for astronomical observation, yet at a reasonable distance from the United Kingdom.

The task of finding such a site has been given to the Royal Observatory Edinburgh, and groups of observers are presently engaged in making systematic observations of seeing, sky transparency, sky brightness and general meteorological conditions at sites in the Canary Islands and Southern Italy. The same type of instrumentation and the same methods of observation are used at all sites. The instrumentation includes polaris trail telescopes constructed at Edinburgh and, mounted next to one of them, a polaris trail telescope placed at our disposal for purposes of intercomparison by Professor M. F. WALKER of Lick Observatory. It is too early to give any final results of this site testing work, but there is every indication that the Canary Islands are likely to provide excellent locations for the establishment of a major observatory.

The exact equipment of the proposed new 'Northern Hemisphere Observatory', its number and size of telescopes, is a matter which is presently very much under discussion. It is envisaged, however, that the equipment will include at least one telescope with an aperture of not less than 150 inches.

Author Index

Subject Index

MIX
Papier aus verantwortungsvollen Quellen
Paper from responsible sources
FSC® C105338

If you have any concerns about our products,
you can contact us on
ProductSafety@springernature.com

In case Publisher is established outside the EU,
the EU authorized representative is:
Springer Nature Customer Service Center GmbH
Europaplatz 3, 69115 Heidelberg, Germany

Printed by Libri Plureos GmbH
in Hamburg, Germany